Production & Management of Fruit Crops in Arid/ Drylands

Production & Management of Fruit Crops in Arid/ Drylands

Editor

Nana Dhumal Deshmukh

scitus
academics

Production & Management of Fruit Crops in Arid/Drylands

Edited by **Nana Dhumal Deshmukh**

ISBN: 978-1-68117-091-6

Library of Congress Control Number: 2015935490

SCITUS Academics LLC,
www.scitusacademics.com
Box No. 4766, 616 Corporate Way,
Suite 2, Valley Cottage,
NY 10989

Notice

Printed in United States of America on Acid Free Paper ∞

Contents

Preface

In every region of the world it is necessary to find or develop appropriate techniques for agriculture. A large part of the surface of the world is arid, characterized as too dry for conventional rain fed agriculture. Yet, millions of people live in such regions, and if current trends in population increase continue, there will soon be millions more. These people must eat, and the wisest course for them is to produce their own food. Yet, the techniques are so varied that only a very large volume would cover the entire subject. This publication is only a primer, an introduction to appropriate techniques. More extensive treatments are mentioned in the bibliography. In many cases the most suitable techniques for a particular region may be those already developed by the local inhabitants. In some cases it will be difficult to improve on local techniques, but at times even simple and inexpensive innovations may be almost revolutionary. This technical note suggests that one must begin to improve local agriculture in arid zones by learning what is already there. Plants are adapted to aridity by several mechanisms. There are plants with a short life cycle that can germinate, grow, and produce during a very short period of available moisture. There are plants with deep or extensive root systems which have the ability to gather water over a wide area. There are plants which store up water in their tissues and release it very slowly. There are plants that are protected from water loss by wax or other impediments. There are plants with very small or narrow leaves, thus reducing water loss. There are plants in which the tissues themselves can withstand much desiccation without dying. Crop plants in arid regions may have any or a combination of such mechanisms.

Editor

Chapter 1

Focus on Agricultural Biotechnology: Prospective for Bio-watersaving Theories and Their Applications in the Semi-arid and Arid Areas

Zhengbin Zhang[1], Hongbo Shao[2, 3, 4], Ping Xu[1], Mengyun Hu[1], Weiyi Song[5], and Xiaojun Hu[1]

[1]Center for Agricultural Resources Research, Institute of Genetics and Developmental Biology, Chinese Academy of Sciences, Shijiazhuang 050021, China

[2]State Key Laboratory of Soil Erosion and Dryland Farming on the Loess Plateau, Institute of Soil and Water Conservation, Chinese Academy of Science and Ministry of Water Resources, Yangling 712100, China

[3]Binzhou University, Binzhou 256603, China

[4]Institute of Life Sciences, Qingdao University of Science and Technology Zhengzhou Road 53, Qingdao 266042, China

[5]Department of Life Science, Shangqiu Normal University, Shangqiu, 476000, China

ABSTRACT

Drought and water shortage have become a world-wide problem in recent years with the effect of global warming. Water-saving agriculture has become an inevitable direction of agronomical research, it has been developed from practices through agronomic water saving to engineering water saving and to the approaches in biological water-saving. The essence of biological water-saving should be the high efficient use of water through biological ways, namely the "utilization and exploitation of physiological and genetic potentials of organisms so as to acquire more agricultural output and better economic and ecological benefits by utilizing smaller or same amounts of water or poor quality water". As a systematic approach, biological water-saving should not only be applied with priority in crop production, but also in other aspects of agriculture and industries such as husbandry, aquaculture, landscaping, sewage water management, water and soil environmental conservations. Therefore, biological water-saving represents the human effort in the construction of a resource-saving and environmental-friendly society.

INTRODUCTION

The twentieth century has witnessed the great effort of mankind to strive for more food production, which has led to an unprecedented increase in grain production known as the First Green Revolution. As mankind entered the 21st century, the Blue Revolution – the Struggle in agriculture for better use of water resources – has just begun. Norman (2000), Nobel Peace Prize Laureate, said in 2000, "how can we continue to expand food production for a growing world population within the parameters of likely water availability? The inevitable conclusion is that mankind in the 21st Century will need to bring about a 'Blue Revolution – more crops for every drop' to complement the 'Green Revolution' of the 20th Century. Water use productivity must be wedded to land use productivity. Science and technology will be called upon to show the way". Water shortage has become an urgent problem in agriculture all over the world. Rapid expansion of cropped land and over-irrigation for high yield have led to the depletion of water resources in many parts of the world, such as the drying-up of lakes and rivers and declining ground water tables. The unsustainable techniques of agricultural development have greatly reduced the water that otherwise could be used for ecological maintenance. If such a water shortage develops into a crisis, it may have effects far more severe than the crisis in oil shortage that we have experienced thus far, or can be expected in the future.

Improving water use efficiency through biological approaches in combination with other water-saving methods and projects and developing a sustainable agriculture under limited water resources have become, and will continue to be the challenges for agricultural scientists in the future (Zhang and Zhang, 2007; Shao et al., 2006-2008). The current article reviews the theories, practices, progress and outlook of bio-water-saving with an emphasis on China.

UNDERSTANDING THE CONCEPT OF WATERSAVING AGRICULTURE IN CHINA

Drought is a severe problem in the 21st century all over the world. Agriculture consumes the largest part of water, about 60-70% of the whole water consumption. How to use the limited water resource efficiently and keep a sustainable agriculture and economy has become a challenging issue for human beings.

From a historical view, China made many achievements in crop drought resistance and water-saving. For example, one of its greatest contribution was Qimin Yaoshu (Essential Techniques for the Peasantry), a monograph in agronomy written between about A.D 530s and 540s. In this great book large information was summarized about dry land-farming technical system with plough, harrowing and leveling as its core parts and drought prevention and moisture retention as its essential ideas, and also recorded upland rice production (Jia, 1963). These drought-resistant and water-saving techniques have been playing an important role in dry land farming in China since ancient time. China also made remarkable achievements in water-saving by hydraulic engineering during the 1950s-1970s, for instance, the construction of many dams and reservoirs and the control of Yellow River and the Huaihe River. In the 1980s, using fertilizers and the improvement of fertilizer performance and water productivity greatly increased crop yields in dry land areas. In the 1990s irrigation using water from rainwater-collecting tanks resulted in high rain water use efficiency. There have been three conceptual developments in dry land farming research in the semiarid regions of China: 1) attention from water deficiency to nutrient deficiency; 2) from water deficiency to the improvement of water productivity; 3) from drought resistance to the high water use efficiency (WUE) breeding (Zhang et al., 2004). The trend to shift from passive drought tolerance research to the active drought resistance and high WUE will inevitably occur in the future (Zhang et al., 2005). Similar research results and conclusions have been obtained in vast semiarid regions including Mediterranean (French et al., 1984), East Africa (Smaling et al., 1992), African regions to the south of Sahara (Rockstrom, 2001), desert steppe of West Africa (Breman et al., 2001), South India (Ahlawat et al., 1998) and West China (Li et al., 2001).

It is believed that water-saving agriculture, proposed based on the concept of water-saving in irrigated agriculture, includes the high-efficient water use in dry land farming. In terms of the cores of high efficiency water use, both water-saving agriculture and dry land farming can be called as high WUE agriculture. High WUE agriculture refers to a new farming system that integrates high-tech system with economy and market and strives for high water use efficiency, high economic, ecological and social returns per unit water consumed (Zhang et al., 2005). For example, many countries like Israel have developed and adopted water-saving irrigation techniques such as water transmission through pipes and drip irrigation to improve water use efficiency on one hand, and produce agricultural products with high added value such as vegetables, fruits and flowers for international market on the other hand so that they have become the backyard of Europe (Yang, 2002). This is a typical high WUE agriculture. Therefore, we can say, water can be used as gold, and soil can be used like gold as well, book knowledge can be used really as gold. That is to say, promoting education and agricultural science and technology and improving farmers' scientific qualities and planting cash crops under limited water conditions can solve the problems of slow agricultural development, rural backwardness and farmers' low income in arid and semi-arid regions of China and in the other developing countries to a certain extent.

NEW TECHNOLOGY IN COMBINATION WITH PLANT BIOLOGY

Deficit irrigation strategies like regulated deficit irrigation or partial root drying have emerged as potential ways to increase water savings in agriculture by allowing crops to withstand mild water stress with no or only marginal decreases of yield and quality.

As a result, improving crop water-use efficiency (WUE) has been a matter of concern to researchers and agronomists in recent years. WUE is discussed either in terms of instantaneous measurement of the efficiency of carbon gain per water loss by plants or as the integral of such an efficiency over time (expressed as the ratio of biomass accumulation or harvested yield to water use) (Zhang et al., 2005). The WUE in the agricultural sector has been slowly improving due to the use of genotypes with increased WUE (Zhang et al., 2005) and due to adoption of innovative cultivation and irrigation practices (e.g. drip irrigation, use of irrigation calendars based on the depth of water table and soil salinity, reuse of wastewater) (Office of Technology Assessment, 1983; Chaves et al., 2003). Drip irrigation, mulching and protected cultivation have contributed to improve WUE in agriculture by significantly reducing runoff and evapotranspiration losses (Shao et al., 2008). Mediterranean countries like Israel or Spain led developments in drip irrigation and cultivation under plastic in the past decades, but China has been strongly investing in these techniques. China has recently emerged as the world's largest producer of greenhouse vegetables and ornamentals (close to 2 million ha) and has about 15 million ha using plastic mulches (Yang et al., 2007). However, the use of drip irrigation remains too restricted suggesting that WUE can still be optimized by adoption of more efficient irrigation practices. Deficit irrigation strategies have the potential to optimize water productivity in horticulture. Nevertheless, the effects of deficit irrigation on yield or harvest quality are crop-specific. Knowledge of how different crops cope with mild water deficits is the basis for a successful application of deficit irrigation into practice (Yang et al., 2007). Rice (Oryza sativa L.) is the most important staple in Asia, providing on average 32% of total calorie intake. To keep up with population growth and income-induced demand for food in most Asian counties, rice production must be increased by 56% over the next 30 years (Karaba et al., 2007). About 75% of total rice production comes from irrigated lowlands. In Asia, irrigated rice accounts for about 50% of the total amount of water diverted for irrigation, which in itself accounts for 80% of the amount of fresh water diverted. Fresh water, however, is becoming increasingly scarce because of population growth, increasing urban and industrial development, and decreasing availability resulting form pollution and resource depletion. Decreasing water availability for agriculture threatens the productivity of the irrigated ecosystem and ways must be sought to save water and increase the water productivity of rice. The high water demand of irrigated lowland rice mainly arises from keeping the field continuously submerged as a result of high evaporation and seepage and percolation in the fields. To reduce water use in irrigated lowland rice, systems of alternate wetting and drying or alternate submergence and nonsubmergence have been developed. It has been reported that alternate wetting and drying systems can maintain or even increase grain yield (Zhang et al., 2005). However, experimental evidence is still scarcely reported in international published reports. Furthermore, there are reports that alternate wetting and drying systems often reduce, rather than increase, grain yield when compared with continuously submerged conditions. Obviously, it remains a major challenge to reduce water input without compromising yield and to optimize scarce water in rice production. For this objective, Yang et al. (2007) established an irrigation system to save water and increase grain yield, therefore enhancing water productivity by proper water management at the field level in irrigated lowland rice. The physiological mechanisms involved were also investigated. In their water-saving irrigation system, limiting values of SWP were proposed as irrigation indices. These values were related to specific growth stages, so that the wetting and drying could meet the growth and development of rice. Our results showed that the technique not only significantly reduced water input, but also increased grain yield, therefore enhancing water productivity. The technique was demonstrated and applied in eight different ecological rice-growing areas in China. Compared with the conventional irrigation that drainage was in mid-season and flooded at other times, the water-saving irrigation technique increased grain yield by 6.1 to 14.2%, reduced irrigation water by 25.4 to 38.2%, and increased water productivity (grain yield per cubic meter of irrigation water) by 26 to 47%. How could this technique save irrigation water as well as increase grain yield and water use efficiency? They also observed that sensitivity of rice to water deficits varied with the growth stages. A mild soil-drying (SWP at –15 kPa) at most growth stages benefited rice plants. The result indicated that keeping the field continuously submerged would be not good for a high yield and much water could be saved from the conventional irrigation. In conclusion, the water-saving irrigation by controlling limiting values of soil water potential related to specific growth stages not only reduced water input, but also increased grain yield, therefore enhancing water productivity. High harvest index, maintenance in photosynthetic rates, reduction in leaf conductance, increases in cytokinin concentrations in roots and activities of key enzymes involved in sucrose-to-starch pathway in the grains may all contribute, at least partly, to high yield and high water use efficiency under the irrigation system. Currently, our group stress on practicing water-saving agriculture to save water in rain-fed agriculture regions and the north China, in particular so as to produce more grain with limited water resources. But actually high WUE farming should be much more practiced in the regions of South China without facing water shortage, that is to say, such farming high-tech such as high-yield and super-high yield plants and animals with high economic return, intercropping, multi-storey farming, greenhouses, film mulching farming, hydroponics and factory-style agricultural production should be adopted in these regions where water is not a factor limiting high grain yields to produce more high-quality agricultural products and achieve better economic, social and ecological returns. Meanwhile, we also place emphasis on high-efficiency exploitation of poor quality water resources by taking the road characterizing high WUE and high efficiency output. In China, grains were traditionally transported from south to north, but since the 1980s when China began its agriculture reform, several large-scale grain transportations from north to south have occurred. This sufficiently proves the grain production capacity has been quickly declining in the south China but risen in north China to some extent. Therefore, we should rethink and reanalyze macroscopic changes in grain production and consumption in China. This will be helpful to understanding the stress on rehabilitating grain production and practicing high WUE farming in humid region, economically well-developed region and the regions that are backward, insufficient and low in grain production. According to theories about virtual water (Allan, 1997; Cheng, 2003), virtual water and soil resources, virtual energy and virtual economy, the problems about interregional loss and gain in economic distribution resulting from the north-to-south grain transfer are practically the problems about inter-regional loss and gain in macro distribution and transfer of such resources as water resources, land resources in production and input energy, and consequently the problem is a problem closely related with sustainable development of water and soil resources and the development of high WUE agriculture in China. The north-to-south grain transfer, transferring insufficient but precious water, soil nutrient, sunshine, warmth and energy resources, and subsidies in the north to the south with sufficient water, light and warmth and a high yield potential, which is commonly known as "the land bestowed with rice and fish" and the "land of abundance". The consequences of such abnormal subsidy and resources transfer have not only severely discouraged the productive enthusiasm in major grain production regions in the north, but also resulted in severe depletion of agricultural water and soil resources in the north, thereby accelerating the crisis of water and soil resources and strangling the potential of sustainable agricultural development. This tendency will inevitably appear in the grain production bases in North China and northeast China plains (Zhang et al., 2005). In China, agricultural development has gone through five major changes: 1) change from the primitive agriculture dominated with labor forces and animal powers to mechanical agriculture; 2) from organic agriculture and then the inorganic and organic combined (with a great deal of straws returned to soil) agriculture to the high output and commercial agriculture; 3) change from grain production-dominated agriculture to dominated by both grain and cash crop production, and to agriculture with well-coordinated grain and cash crops and agro-product production (including animal raising and processing); 4) change from self-sufficient agriculture through domestic market-oriented agriculture to export-oriented agriculture, and 5) change from paddy field-dominated agriculture (needing standing water for a long time) through irrigated agriculture (regular irrigation) and then water-saving agriculture (regularly irrigated at a fixed water amount) to high WUE agriculture in south China, and the change from dry land agriculture through irrigated agriculture and then water-saving agriculture to high WUE agriculture in north China. With rapid development in water-saving agriculture, we have got more in-depth knowledge and understanding of efficient uses of macroscopic water resource and microscopic moisture, formed many new concepts and research directions, and in particular paid more and more attention to the concepts and indexes in raising water resource, water use rate, WUE as well as their related researches. We believe that in China, agriculture has been seen or will see three important stages, "comprehensive development with grain as the key link", "comprehensive development with money as the key link" and "harmonious development with water as the key link". Agriculture has been transforming and will transform from forestry and grassland agriculture through grain-production agriculture, grain-economy agriculture and then economic-grain agriculture to high WUE agriculture and forestry integrated agriculture (Borlaug and Dowswell, 2000; Zhang et al., 2006). From the broad sense of water conservation and high efficient use of water, water and soil conservation may be the fundamental form of the concept of water-saving agriculture or high water / soil use efficiency agriculture in China. It includes more than dry land agriculture, irrigated agriculture and ecological environment improvement and presents an obvious intention to form a new academic discipline. This new discipline not only aims to take biological and engineering measures to fully conserve and make efficient use of water in farm fields and nonfarm fields, but also involves conserving and making efficient use of resources such as soil nutrient; it aims not only to resist drought but also to prevent flood so as to coordinate and improve water resources circulation and ecological environment. Nowadays, the concept of water and soil conservation has been gradually understood and accepted at home and abroad (Wu et al., 2003; Zeng et al., 2005). Biological water-saving shall be the foundation and core of water and soil conservation and of water saving agriculture (Zhang et al., 2007; Shao et al., 2006- 2008). In irrigation regions, water-saving engineering techniques such as canal system of field irrigation and irrigation equipment have been widely extended. So far the irrigated land area has accounted for half of the total cultivated land in China. These have all greatly improved the utilization ratio of river and rainwater resources and hence played an important role in stabilizing grain production in China. Since agricultural and engineering water-saving measures were taken to improve the water resources utilization ratio, and as fresh water resources have been in short supply and polluted and the climate has been becoming drier, it is necessary to employ modern agricultural and new biological techniques to carry out in-depth research on biological water-saving measures to breed and extend new drought-resistant water-saving breeds of living organisms so as to produce more food and more dollars for every drop of water. This is the ultimate way for the development of water-saving agriculture (see the website listed in References). On the condition that China can stably produce 400 to 500 million tons of grain (Lu, 2005), its agriculture should be oriented for stable grain production and profit-making. It can better solve the problems concerned with farmers, countryside and agriculture only by practicing high WUE agriculture and developing export-oriented ago-products with high added value. Under the new situations, the term "green revolution" should mean two things: one is to greatly increase grain yield and the other is to produce more pollutant-free, high-quality and high-price "green foods". We should launch a "blue revolution" centering on utilizing limited water resources at high efficiency in Northwest China and North China that have limited water resources to complete sustainable development of "green revolution" and thus produce more grain. On the other hand, we should launch a "blue revolution" aiming to improve the economic value of water use to produce more pollutant-free, high-quality and high added value "green foods" in South China and Eastern coastal areas Of China that have rich water resources such as river water, ocean water, sea water and slightly brackish water. In the meantime, we should conduct a "white revolution" with milk cow as the key link in the grassland regions and the areas with welled development agriculture (Zhang, 2002). Nowadays as water resources are becoming increasingly insufficient and water quality is deteriorating, agriculture with high water use efficiency is the key for China to develop modern agriculture and join the outside world. Now, China still stresses on developing water-saving agriculture and the first goal of water-saving agriculture is to raise the utilization rate of water resources, which is mainly realized by engineering water-saving from the macroscopic point. However, only engineering and biological measures are combined and then true water-saving can be realized. Thus, the second aim of water-saving agriculture is to increase crop water use efficiency and yields. Although crop water-saving breeding is a newly emerged term, remarkable achievements have been obtained in the past practice of crop drought-resisting and water-saving breeding. The real problem is that we have not previously paid closer attention to it so that the theoretical researches about it have relatively lagged. If it has high agronomic WUE, but no good economic returns, it will not have any developmental hopes. Hence, the third target of water-saving agriculture is to improve its economic return, i.e., economic water use efficiency, while trying to improve its biological water use efficiency. And because plants consume the largest part of water the study of value water use efficiency of plants will gradually attracted more and more concern in the future. The fourth target of water-saving agriculture is to further improve the water use efficiency of animals and microbes through in-depth researches on biological water-saving. The fifth target of water-saving agriculture is to attract people's attention to ecological WUE of plants as the environment problems will become worsen in many regions in the future (Zhang et al., 2004). Agronomic water-saving has developed for thousands of years. Now China is not economically strong enough to practice engineering water-saving in a large area. Living things have a wide genetic diversity thus capable of adapting to various environments, but their exploitation is still not enough. Biological water-saving is simple, easy, cheap to carry out and able to produce good results and high added value. For example, biological products produced under arid conditions are usually good quality, suffered less pest injuries as green foods. Therefore, all agronomic and engineering water-saving measures should be combined and adopted in biological water saving, which will be the ultimate objective of water saving agriculture.

NEW DEVELOPMENT OF THE CONCEPT OF BIOLOGICAL WATER-SAVING

The development of biological water-saving has experienced three stages: water-saving through crop physiology (Shan et al., 1991), water-saving through crop genetics and physiology (Shi, 1999). Water saving trough biological technology includes water-saving by animals, plants and microbes (Zhang et al., 2007; Shao et al., 2007). Scientific development trends in biological water saving science indicate that biological water-saving has become a systematic engineering. The concept of biological water-saving is still expanding. It not only should be given the first priority position in crop production and in particular grain production in agriculture and be placed in an important position in vegetable, fruit and flower productions; but also it has a great water-saving potential through animal breeding. Besides, it has significance in landscape planting, sewage control, water and soil conservation and environment control. Therefore, we think that biological water-saving, or more exactly, biological high efficiency water use, should be defined as "to produce more agricultural output and higher economic and ecological returns by utilizing and exploiting physiological and genetic potentials of living organisms (including animals, plants and microbes) with limited or same amounts of water or poor quality water resources".

Biological Water Resources

Biological water resources refer to the total quantity of water resources contained within the bodies of organisms. As to the estimation of biological water resources, the files of UN Water Conference in 1977 indicated that the global biological water resource reserve is 0.112 (KM 3), accounting for 0.0001 (%) of the total water reserve and 0.003 (%) of fresh water reserve (United Nations, 1977). Hereby, organisms are also a bank of fresh water resources, like forests, grassland and agricultural produce all reserving large quantities of water. Compared with the above mentioned other water resources, biological water resources are especially precious, which are the only source of our food. And water is the blood of the biosphere and the decisive factor of the sustainable development of environment (Malin, 2005).

Precipitation ← Plant more forest and grassland can increase precipitation and improving ecology environment
↓

Ground surface water ← Plant more forest and grassland can conserve more water and soil
↓
Sewage water ← Plants can improve sewage water quality
↓
Biological water ← Plant drought resistance and water-saving crops can save irrigated water and increase water use efficiency
↓
Soil water ← Plant drought resistance and water-saving crops can save soil water and increase water use efficiency
↓
Underground water More plants can promote its sustainable circulation

Figure 1: Bio-water saving and water conversation and conservation.

Although biological water resources are relatively small in comparison with other types of water resources, it has close relationship with the living environment of us human beings. Biological water is one of the central joints in the conversion of various water resources. Therefore, there are still large research space and development potentials for biological water.

The Direction of Biological Water-saving

The scarcity of freshwater is a difficult problem in the world. It is the common choice for the countries to develop water-saving society and improve the higher efficient use of water resource. The improvement of plant intrinsic WUE is one of the important connotations for agricultural water-saving except engineering water-saving, agronomical water-saving and management water-saving. Furthermore, this is the key and potential for the further improvement in water-saving and production increase (Shan et al., 2006). We consider that biological water-saving includes three levels of implication in agronomic term: one is to reduce water and soil loss by organism mulching; the second is to reduce irrigation and soil water depletion, namely, water-saving; the third is to improve WUE so as to produce more food and economic benefits by every drop.

STUDIES ON THE TRANSFORMING SYSTEM OF SIX KINDS OF WATER RESOURCES WITH BIOLOGICAL WATER TRANSFORMATION AS THE CORE

Before 1980s, many studies focused on "transformation of three kinds of water" (rainwater ground surface waterunderground water); in 1990s more research focused on "transformation of four kinds of water" (rainwaterground surface watersoil water underground water) (Liu, 1994); in 2000s more studies focused on "transformation of five kinds of water" (rainwaterground surface watersoil waterplant water underground water) (Kang et al, 1994; Wei et al., 2000). We think that now the research should be carried out on the "transformation of six kinds of water" (the precipitation—ground surface water—sewage water—biological water—soil water—underground water) (Figure 1). All kinds of human activities and industrial and agricultural productions change clean water into vast amounts of sewages in rural and urbane areas. As a result, Sewages have become a resource that must be recycled and then used. Therefore, sewage disposal has become a world-wide problem (Bao, 2005). Living things can not only conserve water and soil, but also decontaminate sewage, for example, sewage water can be cleared by many plants like reed and other animals and microorganisms. Therefore, biological water-saving can play an important role in the transformation of six kinds of water. Water resources are flowing resources and act in a cyclic system. Thus, we must strengthen the researches about timely and highly efficient water use by living things. If water is not timely and fully used by them, that means water extravagance. If there was a reservoir carrying no living things, then its water would not be ecologically dead; if it carries living things, it can be used for fish raising and irrigation, and then its water is ecologically and economically active. It is a future development trend to make use of living things to improve poor quality water resources. Academics and oceanographer in China (Zeng, 2000) advocated "blue revolution" and "farming fishes in sea", namely seeking foods from sea. In this aspect, there is a high potential in high-efficiency water use to be exploited in the coastal areas of China.

In 1980s, some famous Chinese scientists, represented by Qian (2000) brought forward "sand industry", which is a good example in high efficiency biological water use (Liu, 1995). In deserts and newly reclaimed lands, autogenous wild grasses firstly appear, then certain ecological effects occur and animal raising creates economic benefits later on. Although plants are major consumers of water resources, plants can bring about great ecological and economic benefits through water consumption and transformation. In the desert part of the semi-arid area in China, Psammophytes along with other various approaches to efficiently utilize limited rainwater and soil water should be greatly expanded in order to improve ecological environments for human welfare. There are many people who have emerged as the desert-controlling paradigms in many regions of Shaanxi, Gansu, and Ningxia provinces in China. In wild deserts, they have planted trees that form ecological forests that do not have previous examples in history and the scientists of the institution of botany, Chinese Academy of Sciences are making effort to improve and construct grasslands through natural recovery in the grassland region of Inner Mongolia, China. These have attracted high concern in the desert-controlling field of the world and the government of China. In such arid areas, such as Qinhai and Lanzhou of Northwest China, the annual rainfall is only 100-200 mm, and the annual evaporation is at least 2-5 times the rainfall, amounting to 1000 mm; but, the farmers still grow crops on sand-mulched lands, for instance, seed-use watermelon, which achieving a good economic return, and this fully demonstrates that high-efficiency biological water use is feasible, realistic and important under limited water conditions. It follows that so long as desert control is scientifically conducted it will produce the effects that plants expand while deserts retreat so as to improve desert environments and help desert-related economic development (Zhu, 2004), just as what the examples in high efficiency water use has shown.

TO RESTRUCTURE CROP PRODUCTION AIMING AT BIOLOGICAL WATER-SAVING

Although fallow can reduce nutrient and soil water consumption, but under certain circumstances, planting and burying green manure plants in soil can increase soil fertility, maintain fertility, or planting such minor crops as legume, buckwheat and oil sunflower can also result high efficiency water use, certain economic return. The practices to plant such cold-resistant crops as spinach, Brassica chinensis, and B. chinensis var. oleifera on fallow lands in central Shaanxi of China and the regions of South China in autumn and winter are the examples in high efficiency water use. In humid regions and seasonally dry regions of South China such drought-tolerant crops as upland rice, maize, sorghum and wheat can be planted in the place of rice and potato can be planted on autumn- and winter-fallowed lands after rice harvesting (Lin, 2005), which is an approach in high-efficiency biological water use. In cold zone of north China, many regions have been conducting the studies about expanding winter wheat northward because of global warming (Gao et al., 2005) in order to utilize soil water and other resources in autumn and winter, and to conserve water and soil, reduce sandstorm occurrence and improve ecological environment. Therefore, restructuring the distribution and planting systems of different plants and breeding and extending water-saving drought-resistant plant varieties will make biological water-saving and high-efficiency water use have great potentials to be exploited.

PHYSIOLOGICAL REGULATION AND GENETIC MODIFICATION TO IMPROVE CROP WUE

Our long-term research has indicated that drought resistance and water-saving is the means for crop to survive and reproduce under arid and semi-arid climatic environments, and the goals in crop production are high WUE and high yield (Chaves et al., 2003; Karaba et al., 2007;

Shao et al., 2005-2008). Although drought resistance and water-saving are complex problems in plants, human beings have still made many remarkable achievements in crop breeding and production and will continue to advance in these respects. In semi-arid areas, drought-resistance, water-saving and high yield are not in severe conflict under some condition, some drought-resistant crop varieties with high yield and good quality; and super-high yield varieties can be bred to raise the efficiencies of water, solar energy and nutrients. WUE can integrate drought-tolerance and high yield in one body (Zhang, 1998) and is a quantificational comprehensive index in drought-resistance and water-saving research, including many aspects such as drought resistance, drought-tolerance, water-saving high efficiency water use and high-efficiency output; now it have gradually been becoming an important indicator to assess crop drought resistance and water-saving (Zhang, 2003, 2006; Shi, 2005; Uri, 2005; Su et al., 2007). In recent years, we have carried in-depth studies about the physiology, genetics, breeding, evolution, gene mapping and molecular marker of wheat in WUE (Zhang et al., 2000; Zhang et al., 2002), showing that wheat has increased the WUE of its leaves, individual plants and population through evolution to some extent (Zhang et al., 1998) and its WUE-related traits are genetically controlled are possible to be genetically and culturally improved. Under arid and semi-arid conditions, the WUE of wheat ranges between 0.1 and 1 kg/mm, and the WUE of maize ranges from 0.2 to 2 kg/mm, and as a result the WUE difference between the two is tenfold. On condition that the annual rainfall is 500 mm, wheat may have a yield of 7,500 kg/h and maize a yield of 15,000 kg/ha. This suggests that the concept of "one kilogram water for one kilogram grain" has been outdated by variety improvement, farming and fertilization practices and so on. Malse et al. (2005) of Australia published a paper in Nature which told that they adopted 13C as the representative trait of WUE, then positioned the QTL controlling transpiration efficiency on the ERECTA marker of the second chromosome and thereafter cloned from Arabidopsis ERECTA gene, a gene encoding a putative leucine-rich repeat receptor-like kinase that can change both leaf stomatal number and leaf structure, and regulate the flowering time, and is proved to regulate plant transpiration efficiency and consequently to have a bright prospect in improving crop drought resistance and using water at high efficiency. Some people call living things that have a high water requirement as "water pigs", like as buffalo, rice and clover; living things that have a low water requirement as "water misers", like as camel, cactus and millet. Water pigs can be transformed into water misers through natural evolution and artificial improvement (Perry et al., 2001). For example, rice can be transformed into upland rice and Shanghai Agricultural Gene Center of China bred hybrid upland rice characterizing drought-resistance, water-saving, high quality and good quality (Yu et al., 2005). In particular, modern molecular, genetic and transgenic techniques can be employed to transfer drought resistance and water-saving genes among different species to breed new varieties of living things with drought resistance, water-saving, good quality and high yield, which represent an important research direction in biological water-saving. In 2000, Elumalai, an American scientist, reported that HVA1 gene of barley was transformed into wheat and the offspring of the transgenic wheat got their WUE and drought-resistance greatly-improved (Chaves et al., 2003; Karaba et al., 2007; Shao et al., 2007, 2008). Therefore, there exists a great potential to be exploited in improving the use efficiencies of soil water, rainwater and irrigation water through variety and soil fertility improvements and farming practice optimization. With their long-term efforts the scientists and experts in crop genetics, breeding and production may be able to raise the maximum yield potentials of such C3 plants as rice and wheat to caught up with and even surpass the yields of some varieties of such C4 plants as maize and sorghum, which may fully demonstrate the potentials and importance of WUE improvement in crops. On the production level of 9,000 kg/ha, the field WUE of wheat (grain yield/ unit water consumed) is higher than that of maize and then that of rice because different crops have different water requirements. The developmental trends of dry land crops with high yields reveal that in China wheat yield is 2 250 kg/ha on poor dry land making up 50% of its total dry land, 3 750 kg/ha on mid-yield dry land making up 20-30% of its total dry land, 6 000 kg/ha on fertile dry land accounting for about 15% of its total dry land and about 7500 kg/ha on breeding land and demonstration land making up the rest 5% of its total dry land. In the region of China where the rainfall stands at 500 mm, the average yield of wheat on dry land did not reach 750 kg/ha in 1950 and now ranges within 2250-3750 kg/ha. The yield of wheat on dry land reaches 7500 kg/ha in arid area of Laizhou, Shandong province (Lin, 2005), showing that there is great potentials to be exploited in high-yield breeding and production of dry land crops. In addition, with water-saving and drought-resistant varieties with high yields extended and adopted the crop yield still can reach 7500 kg/ha even though hybrid rice is planted in the place of rice, wheat adaptable to both dry land and irrigation land is planted in the place of wheat adaptable to irrigation land or irrigation is conducted 1-3 times instead of 5-10 fives during growing period. In particular, the successful breeding, extension and adoption of maize, rice and wheat varieties with super-high yields, that is, raising water use efficiency by improving crops, are of important significance and indicate a great potential to be exploited in increasing grain yields in China. It is shown that in China great progresses have been made in dry land wheat breeding on the one hand and there exists a great potential to be exploited in high-yield breeding and culture of dry land crop on the other hand. It is our belief that in China, as the conditions in wheat production is improved and the fertilizer input in wheat production is increased, wheat yield will averagely reach 3 000 kg/ha or more on dry land and even 3 750 kg/ha on 50% of the dry land in the middle of this century. Therefore, raising WUE by improving crop is of important significances in continuous increasing the grain yield of China brought to the persistent increase in China's food production. Plants genetically and physiologically differ significantly in drought resistance and high water use efficiency, just as "someone can gain weight even by drinking water". Wheat and other crops can be classified into 4 WUE types of which type 1 is characterized by low water consumption and low yield and low WUE, and consequently referred to as so-called "three-low" type; type 2 is characterized by moderate to low water consumption, middle to high yield and higher WUE, for instance, Xifeng 20, Chang 6878 that are planted in the region of winter wheat in north China; type 3 is characterized by medium to high water consumption, the highest yield and the highest WUE, such as Shijiazhuang 8 and Linfeng 615 that are the varieties of wheat adaptable to both dry land and irrigation land that are extended in the dry land area of Huanghuaihai Region in China; and type 4 is characterized by high water consumption, low yield and low WUE. We hope to breed new drought-resistant and water saving varieties with high yields that belong to types 2 and 3 (Zhang et al., 2006). Considering genetic improvement of drought-resistance and water-saving crop varieties, the three measures can be taken as follows: 1) to strengthen the screening and exploitation of various drought-resistant germ plasms with high WUE; 2) to introducing and domesticating new plant species and crop varieties with drought resistance and high WUE from arid and semi-arid areas of China and foreign countries; 3) to breeding new drought-resistant and water-saving crop varieties with high-quality and high yield by employing biotech including conventional breeding methods and transgenic techniques. Now in-depth research on the physiological and genetic mechanisms of high WUE crops should be carried out on the levels of growth, development, morphology, structure, tissue, cell, molecule, gene metabolic regulation, genome, genetic engineering, breeding and improvement in the following aspects: (1) drought resistant mechanisms; (2) high WUE mechanisms, such as molecular biology mechanisms in aquaporin and WUE regulations; (3) the mechanisms to increase the harvesting indexes (HI) of crops; (4) the coordination among drought resistance and high WUE and high yield; (5) the mechanisms to couple high use efficiencies of water and nutrients; (6) high efficient use and transport of water and light. Thus, biological water-saving should be stressed in future. The research about biological water-saving need to be reinforced and a project to study the mechanism and application of biological water-saving or the mechanism and application of the improvement in crop WUE should be organized, which will be of important significance to the developments of water-saving agriculture and sustainable agriculture.

CONCLUSIONS

In nature, plants also have to cope with the interaction of multiple stresses that often arise concomitantly with drought, and ultimately involve oxidative stress (Stanhill, 1992; Richards et al., 2002; Li and Peng, 2003; Inter Academy Council, 2004; Liang, 2006; Shao et al., 2008). Protective responses at the leaf level must then be triggered in response to stress to prevent irreversible damage to the photosynthetic machinery. The molecular understanding of mechanisms involving stress perception, signal transduction, and transcriptional regulation of stress tolerance, may help engineer tolerance to multiple stresses in crop plants. Application of TFs may provide a more adaptive function in improving tolerance, either through protection or repair mechanisms. Advances in the molecular biology of stress responses in tolerant organisms are introducing the potentials of stress tolerance genes in agricultural programmes, not only to ensure survival but also to ensure productivity under drought environments. The researches about biological water-saving contain wide ranges of diverse contents; have a weak basis and later starting so that they have a great potential. The technology of biological water-saving will be the ultimate aim in saving water and increasing crop yields. It also can be seen that biological water-saving is the ultimate aim not only in saving water in agriculture but also in constructing a resource-saving and environment-friendly society. Bio-water-saving has been listed as one of the national key research fields of China and one of the key research fields in which related organizations and some provinces of China hope to make breakthrough, but although biological water-saving is in the ascendant, it will have a long way to go as well. It is believed that it is the main task to construct and expand the agriculture characterized by biological water saving as well as construct and promote the cities characterized by biological water-saving and it is the ultimate goal to construct a society by biological water saving. The construction of the agriculture characterized by biological water-saving involves two respects: one is crop planting characterized by biological water-saving and the other is animal raising characterized by biological water-saving. Crop planting characterized by biological water-saving is the main research field to be studied, whereas it is imperative to study animal raising characterized by biological water-saving. Crop planting characterized by biological water-saving is the core of water saving agriculture ranking the highest in importance. As to crop planting characterized by biological water-saving the first is to construct and expand field crop planting characterized by biological water-saving and the second is to construct and expand vegetable crop planting characterized by biological water-saving. As urban and township developments are ongoing quickly, the problem about the water consumption of landscape planting and magnitude of sewage disposal has attracted the attention from all the society and become one of the key problems critical to urbane construction and sustainable development. Therefore, the two keys to construct and promote the cities characterized by biological water-saving are to construct and promote landscape planting, sewage disposal and environment conservation characterized by biological water-saving. It would have significant reality meaning and long history influences that will have realistic significance and profound historical influence. Therefore, efforts should be made to transforming landscape planting characterized by high water consumption into landscape planting characterized by water-saving as early as possible so that urbane landscape planting will embark on a sustainable and healthy road. Biological water-saving is a systematical engineering that can play an important role at the different parts of the ecosystems. Besides the agriculture and cities characterized by biological water-saving will be constructed, the techniques of biological water-saving should be employed to reinforce the environment control and improvement and the development and exploitation of poor quality land and water resources for water and soil conservation in particular in non-farming and mountainous areas so as to construct a ecosystem and a society both of which will be characterized by biological water-saving.

ACKNOWLEDGMENTS

This work is jointly supported by the Hi-Tech Research and Development (863) Program of China (2006AA100- 201), National Science and Technology Supporting Plan of China (2006BAD29B02), the Major Project of Chinese Academy of Sciences (KSCX1-YW-09-07) (To Zhang ZB) and the Cooperative and Instructive Foundation and Open Foundation of State Key Laboratory of Soil Erosion and Dryland Farming on the Loess Plateau (10501- HZ; 10501-194) (To Shao HB).

REFERENCES

1. Ahlawat IPS, Rana DS (1998). Agro-techniques. In: Yadav RL, Singh P, Prasad R, Ahlawat IPS. Fifty years of agronomic research in India. New Delhi: Indian Society of Agriculture.
2. Allan JA (1997). Virtual water: A long term solution for water short Middle Easterm economics? Brithish Association Festival of Science, Unversity of Leeds, September
3. Bao XR (2005). A study on changing sewage into resource in China. Environ. Sci. Technol. 2: 51-53.
4. Breman H, Groot JJR, Keulen H (2001). Resource limitations in Sahelian agriculture. Global Environ. Change, 11: 59-68.
5. Chaves MM, Maroco JP, Pereina JS (2003). Understanding plant responses to drought-from genes to the whole plant. Functional

Plant Biol. 30: 239-264. Cheng GD (2003). Virtual water-A strategic instrument to achieve water security, Bull. Chin. Acad. Sci. 18(4): 260-265.

6. Elumalai S, Bahieldinl A, Wraith JM, Wener LM, John PL (2000). Improved biomass productivity and water use efficiency under water deficit conditions in transgenic wheat constitutively expressing the barely HVA1 gene. Plant Sci. 155: 1-9.
7. French RJ, Schultz JE (1984). Water use efficiency of wheat in a Mediterranean-type environment. I. The relation between yield, water use and climate. Australian J. Agric. Res. 35: 743-764.
8. Gao ZQ, Miao GY, Deng ZF (2004). Global climate change and winter wheat planting in northern region. Rev. China Agric. Sci. Technol. 1: 9-13.
9. Inter Academy Council (IAC) (2004). Realizing the promise and potential of African agriculture: Science and technology strategies for improving agricultural. Published by the InterAcademy Council in June 2004. Amsterdam, Netherlands. Improving durum wheat for water use efficiency and yield stability through physiological and molecular approaches. http://www.medaqua.org/forum/IDUWUE.html.
10. Improving water use efficiency in Mediterranean agriculture. http://www.distagenomics.unibo.it/wuemed/index.html.
11. Jia S (1963). Qiqiminyaoshu. Beijing, Agriculture Publishing House, pp. 66-89.
12. Kang SZ, Liu XM, Xiong YZ (1994). Theory and application on soilplant-atmosphere continuous body water transmission. Beijing, China Water Power Press.
13. Li TS, Peng SQ (2003). Dry farming and water-saving agriculture in Egypt. World Agric. 4: 40-42.
14. Li XY, Gong JD, Gao QZ, Li FR (2001). Incorporation of ridge and furrow method of rainfall harvesting with mulching for crop production under semiarid conditions. Agric. Water Manage. 50: 173-183.
15. Liang BZ (2006). Six great strategic missions of Chinese scientific and technological innovation in future 15 years by agricultural department proposed. http://www.agri.gov.cn/xxlb/t20060706_643710.htm

16. Lin Q (2005). Water-saving and high-yield culture by early and deep fertilizing in dryland wheat. http://www.longread.net/read.php?tid =4799.
17. Lin GB (2005). The key technology on rice-maize-autumn potato–winter potato cultivation model. Fujian Agric. Sci. Technol. 4: 64-65.
18. Lu LS (2005). New trend of agricultural development and food security in China. Eng. Sci. 4: 30-35.
19. Liu CM (1994). Study prospect and expectation in 21 century of Geographical hydrology. J. Geography, pp. 7-10.
20. Karaba A, Dixit S, Greco R, Aharoni A, Trijatmiko KR, Marsch-Martinez N, Krishnan A, Nataraja KN, Udayakumar M, Pereira A (2007). Improvement of water use efficiency in rice by expression of HARDY an Arabidopsis drought and salt tolerance gene. Proceed. National Acad. Sci. 104: 15270-15275.
21. Malin F (2005). A road to sustainable city by fresh water resourse. Transition to sustainable in the 21st century–The contribution of science and technology. Edited by InternAcademy Panel on Internation Issues. Translated by Fu Bojie, Liu Jian, Cao Jinghua, etc. Beijing: Sci. Pub. pp. 61-65.
22. Masle J, Gilmore SR, Farquhar GD (2005). The ERECTA gene regulates plant transpiration efficiency in Arabidopsis. Nat. 10: 1-5.
23. Norman E, Borlaug, Christoper D (2000) Global food security: Hardness science in the 21st century, Gene Technology Forum, Kasetsart University, Thailand, March 7, 2000.
24. http://bangkok.usembassy.gov/services/irc/gmo29.htm.
25. Office of Technology Assessment (1983) Water-related technologies for sustainable agriculture in U.S. arid/semiarid Lands. Washington, DC: US Congress, OTA-F-212, and October. http://www.wws.princeton.edu/~ota/disk3/1983/8331_n.html.
26. Perry M (2001). Cropping Systems, Water Misers and Water Pigs: Being a Better Bettor, http://weeds.montana.edu/systems/water_article.htm, 07/09/2001
27. Qian XS (2000). Western development on the based of agricultural development. Territorial Economy, 4: 10-11.

28. Rockstrom J (2001). Green water security for the food makers of tomorrow: windows of opportunity in drought-prone savannahs. Water Sci. Technol. 43: 71-78
29. Richards RA, Rebetzke GJ, Condon AG, Herwaarden AF (2002). Breeding opportunities for increasing the efficiency of water use and crop yield in temperate cereals. Crop Sci. 42: 111-121.
30. Shan L, Deng XP, Zhang SQ (2006). Advance in Biological water-saving Research: Challenge and Perspectives. Sci. Foundation China, 20(2): 66-71
31. Stanhill G (1992). Irrigation in Israel: past achievements, present challenges, and feature possibilities. In: Shalheveret J, Liu Changming and Xu Yuexian (eds), Water Use Efficiency in Agriculture. Priel Publishers. pp. 63-77.
32. Smaling EMA, Nandwa SM, Prestels H, Rotter R, Muchena FN (1992). Yield response of maize to fertilizers and manure under different agro-ecological conditions in Kenya. Agric. Ecosyst. Environ. 41: 241- 252.
33. Shi YC (2005). Issues problems on developing Biomass industry. China Basic Sci. 6: 3 6.
34. Shan L, Xu M (1991). Agriculture with water-saving use and the basis of physiology and ecology. Chinese J. Appl. Ecol. 2(1): 70-76.
35. Shi YC (1999). Biological water-saving is the new way to exploit. Sci. Technol. Rev. 10: 3-5.
36. Shao HB, Liang ZS, Shao MA (2006). Osmotic regulation of 10 wheat (Triticum aestivum L.) genotypes at soil water deficits. Biointerfaces 47: 132-139.
37. Shao HB, Chen XY, Chu LY, Yuan YB, Zhao CX, Hu ZM (2006). Investigation on the relationship of Proline with wheat anti-drought under soil water deficits. Biointerfaces, 53: 113-119.
38. Shao HB, Guo QJ, Chu LY, Hu YC, Cheng JF (2007). Understanding molecular mechanism of higher plant plasticity under abiotic stress. Biointerfaces, 54: 37-45.
39. Shao HB, Jiang SY, Li FM, Chu LY, Shao MA, Li F (2007). Some advances in plant stress physiology and their implications in the systems biology era. Biointerfaces, 54: 33-36.

40. Shao HB, Chu LY, Shao MA (2008). Calcium as a versatile Plant Signal Transducer under Soil Water Stress. BioEssays, 30: 634-641.

41. Shao HB, Chu LY, Shao MA, Zhao CX (2008). Advances in functional regulation mechanisms of plant aquaporins: Their diversity, gene expression, localization, structure and roles in plant soil-water relations Review. Mol. Membrane Biol. 25: 1-12.

42. Shao HB, Chu LY, Lu ZH, Kang CM (2008). Primary antioxidant free radical scavenging and redox signaling pathways in higher plant cells. Int. J. Biological Sci. 4: 8-14.

43. Su ZY, Zhang JS, Wu WL, Cai DX, Lv JJ, Jiang GH, Huang J, Gao J, Hartmann R, Gabriels D (2007). Effects of conservation tillage practices on winter wheat water-use efficiency and crop yield on the Loess Plateau,China. Agric. Water Manag. 87: 307-314.

44. The biology of transpiration: From guard cell to globe. http://www.aspb.org/meetings/transpiration06/.

45. The Generation Challenge Programme. http://www.generationcp.org/index.php.

46. The genomics of plant water use efficiency project. http://isotope.bti.cornell.edu/home.html.

47. The world life sciences forum biovision. http://biovision.pharmabc.com/.

48. United Nations (1977) Water Development and Management. Proceedings of the United Nations water Conference, Part4, Programon Press, Oxford, 1978.

49. Uri S (2005). Sustainable development of water recourse. Transition to Sustainable in the 21st century – The contribution of science and technology. Edited by InternAcademy Panel on Internation Issues. Translated by Fu Bojie, Liu Jian, Cao Jinghua,etc. Beijing: Science Publisher. pp. 61-65.

50. Zeng CK (2000). Blue agriculture is developed by aquatic product in the direction of farming and ranching. Sci. Manag. 4: 11-13.

51. Wei C, Steudle E, Tyree MT (2000). Reply: Water ascent in plants. Trends Plant Sci. 5: 146-147.

52. Wu PT, Wang YK, Feng H, Fan XK, Gao JE (2003). Innovation and development of soil and water conservation science of China in 21st century. Sci. Soil Water Conservation, 2: 84-87.

53. Yang RW (2002). Water-saving agriculture in Israel. World Agric. 6: 32- 34.
54. Yu XQ, Mei HW, Li MS, Liang WB, Xu XY, Zhang JF, Luo LJ (2005). Development and application of water-saving or drought tolerant Hybrid rice. Mol. Plant Breed. 5: 637-641.
55. Yang JC, Liu K, Wang ZQ, Du Y, Zhang JH (2007). Water-Saving and High-Yielding Irrigation for Lowland Rice by Controlling Limiting Values of Soil Water Potential J. Integrative Plant Biol. 49: 1445-1454.
56. Zeng DL, Lu SG, Yan PH (2005). Some experiences after attended the 13th international Soil and Water Conservation conference. Soil Water Conservation China, 8: 4-5.
57. Zhang ZB, Xu P, Cui YT, Chen ZB, Liu MY (2004). Three Opinions Changed in Dry-land Agriculture. Rev. China Agric. Sci. Technol. 2: 16-19.
58. Zhang ZB, Xu P, Dong BD, Liu MY, Duan ZY, Liu B, Zhu YG (2005). Water use efficiency is the key research direction in the future agriculture. World Sci. Tech. Res. Dev. 1: 52-61.
59. Zhang ZB (1998). Discuss on the problem of water use efficiency in wheat. J. Tritical Crops, 1: 35-38.
60. Zhang ZB, Xu P, Zhou XG, Dong BD (2006). Advance in genetics improvement of water use efficiency in Crops. Scientia Agicultura Sinica, 39(2): 289-294.
61. Zhang ZB, Shan L, Xu Q (2000). Background analysis of genes controlling water use efficiency of Triticum. Acta Genet. Sinica, 3: 240-246.
62. Zhang ZB, Shan L (1998). Comparison study on water use efficiency of wheat flag leaf. Chinese Sci. Bull. 43(14): 1205-1210.
63. Zhang ZB, Shao HB, Xu P, Chu LY, Lu ZH, Tian JY (2007). On evolution and perspectives of bio-watersaving. Colloids and Surfaces B: Biointerfaces, 55(1): 1-9
64. Zhang ZB, Xu P (2002). Review on study of wheat genome. Hereditas (Beijing), 3: 389-394.
65. Zhang ZB (2006). Development and research of dryland and high WUE agriculture in China. Chinese Science Publishing, Beijing. pp. 45-49.

66. Zhang ZB, Xu P, Dong BD, Chen ZB, Duan ZY, Liu B, Zhu YG, Wang DS, Sun YX (2005). High WUE agriculture is the forward way for Chinese agriculture developing. World Sci. Tech. Res. Dev. 2: 63-71.

67. Zhang ZB, Zhang JH (2007). Water-saving Agriculture: An Urgent Issue. J. Integrative Plant Biol. 49(10): p. 1409.

68. Zhang LX (2002). The call of white revolution. China Dairy. 1: p. 5.

69. Zhang ZB (2003). Fundamental of physiology and genetics and breeding in crop drought resistance and water-saving. Chinese Science Publishing, Beijing: pp. 32-38.

70. Zhang ZB, Liu MY, Zhong GC, Yuan P (2001). Putting more attention to biotechnology in biological water-saving. Bull. Chin. Acad. Sci. 4: 292-294.

71. Zhu JF (2004). Chinese sand industry. China Forestry Press. pp. 8-16.

Chapter 2

Engaging Smallholder Farmers with Seasonal Climate Forecasts for Sustainable Crop Production in Semi-arid Areas of Zimbabwe

Chagonda[1], F. T. Mugabe[2], A. Munodawafa[1], C. P. Mubaya[2], P. Masere[1], and C. Murewi[1]

[1]Faculty of Natural Resources Management and Agriculture, Midlands State University, Bag 9055, Gweru, Zimbabwe.

[2]Directorate of Research and Resource Mobilisation, Chinhoyi University of Technology, Private Bag 7724, Chinhoyi, Zimbabwe.

ABSTRACT

Small holder farmers in the semi-arid areas have not been able to utilize seasonal climate forecasts in making crop management decisions due to limited exposure as well as failure to interpret it. Three participatory meetings were held with farmers in Lower Gweru and Lupine Districts of Zimbabwe soon after the release of the 2008/2009 and 2009/2010 seasonal climate forecast in September of each season. This was done to solicit for farmers' prediction of the coming rainy season and come up with field test crop management practices. Farmers have their sets of indicators that they rely on to forecast seasonal rainfall which are based on generations of past experience that include environmental, biological, and traditional beliefs. There is however, need to investigate how their indicators compare with the scientific forecast in making farming decisions. Farmers claimed that the rains were starting later and finishing earlier whilst analysis of the rainfall record showed that starting rains have been later by 5 to 10 days in the last five years. There was an increase in the number of dry spells in the rainy season in the last five years compared to the period 1980 to 2008. Adding recommended amounts of fertilizers resulted in yield increases of 40% compared to adding half the recommended amounts. Growing long season varieties resulted in higher yields (22%) than growing short season varieties in the 2009/2010 wetter season while growing of shorter season variety had yield advantage (36%) over the long season variety in the 2009/2010 which was relatively dry. Frequent weeding resulted in about 8% increase in maize yields compared to weeding once in a season. The study demonstrates that knowledge of the coming season assists smallholder farmers in coming up with adaptive strategies for climate variability and change.

INTRODUCTION

The severity, duration and frequency of droughts especially in Sub Saharian Africas' semi-arid areas is expected to increase due to climate change (Lasage et al., 2008). Climate variability and change directly affects agricultural production, as agriculture is inherently sensitive to climatic conditions and is one of the most vulnerable sectors to the risks and impacts of global climate change (Parry et al., 1999). Among the most vulnerable groups to the effects of climate variability and change are the small holder farmers who live in semi-arid dry communal areas and have the least ability to adapt to climate variability and change (IPCC, 2001, 2007). This makes smallholder farmers living in these marginal areas food insecure especially now with changing climate expecting to increase the frequency of droughts (Kalanda-Joshua et al., 2011). This implies that farmers have to take a number of crucial crop management decisions and land and water management decisions before and during the growing season, based on climate information (Sivakumar, 2006). It is therefore important for farmers to integrate the issues of climate variability into resource use and development decisions for a more informed choice of policies, practices and technologies that may reduce agriculture's long term vulnerability to climate variability and change (Sivakumar, 2006).

The introduction of seasonal climate forecast into management decisions can reduce the vulnerability of agriculture to droughts (Patt and Gwata, 2002) and reduce negative impacts. It is widely accepted that climate prediction information should be introduced into the planning process as an input into the design of adaptation/ mitigation plans (Zievogel and Zermoglio, 2009; Sivakumar, 2006), and this should be embedded into existing strategies of individuals and communities including their use of indigenous knowledge systems in weather forecast (Prowse and Scott, 2008; Bharara and Seeland, 1994). For seasonal climate forecasts (SCF) to have an impact on decision making, they have to be reliable, timely, and appropriate in context and readily accessible (Challinor, 2008; McCrea et al., 2005). A study carried out by McCrea et al. (2005) shows that adoption and use of seasonal climate forecasts by farmers depends on the level of understanding of the forecast, the format of presentation of the forecast and the attitude of farmers towards the usefulness of the forecast as an indicator of future rainfall.

Seasonal climate forecasts have been used before to assist farmers in making appropriate crop management decisions (Patt and Gwata 2002; Carberry et al., 2002; Meinke and Hochman, 2000) including crop choice, choice of cultivar, fertiliser use, area planted to a given crop, timing and tillage type. Farmers who attended participatory training workshops on forecasts are significantly more likely to use the forecasts than those who learn of the forecasts through non-participatory channels (Ziervogel and Calder, 2003).

Some studies in Zimbabwe (Patt et al, 2005) and Brazil (Ziervogel and Calder, 2003), demonstrate that farmers who used seasonal climate forecast significantly improved their yields compared to their counterparts who did not use it.

Chagonda et al. 669

There are several case studies where farmers have been engaged with improved management of risks under variable climate, but there has not been any quantitative evidence. Most studies report on how farmers who have undergone training on the use of SCF understand better on how to respond to SCF compared to their counterparts who did not undergo the training (Patt et al., 2005). Challinor (2008) argues that the most effective way to conduct pro-poor adaptation research may well be to take (from the onset) a holistic view that is informed by engagement and partnership with potential benefactors. Such projects should integrate indigenous knowledge systems and traditions of the benefactors (Nyong et al., 2007; Nyong and Kanaroglou, 1999). Rainfall characteristics such as onset, ending, dry spells, rainfall intensity and rainfall amounts are showing signs of change in Zimbabwe. The smallholder farmers in Zimbabwe depend on rainfed agriculture which is vulnerable to low, erratic and variable rainfall. Decisions on crop management strategies are very important in such semi-arid areas and are potentially improved by seasonal climate forecast. The objectives of the study were to (i) compare and document farmers' perceptions of changing climatic conditions with measured weather data, (ii) compare climate predictions obtained from the scientific seasonal forecast with the indigenous seasonal prediction and (iii) come up with management options that respond to the seasonal predictions and compare the different options in field experiments.

RESEARCH METHODS

Study Areas

The research was carried out in the 2008/2009 and 2009/2010 farming seasons in Lower Gweru and Lupane Districts in Southwestern Zimbabwe (Figure 1) which lie in Natural Region IV. In Lower Gweru, district, Nyama and Mudubiwa wards were chosen while in Lupane, Daluka nad Menyezwa wards were chosen. The sites were chosen on the basis that they are both in the semi-arid regions of the country dorminated by smallholder farmers. The natural regions are a classification of the agricultural potential of the country, from natural region I, which represents the high altitude wet areas to natural region V which receives low and erratic rainfall averaging 550 mm per annum (Vincent and Thomas, 1960). These two districts are located in the semi-arid areas which cover 75% of the country. These districts have an annual rainfall of less than 700 mm per year and a growing season of 90 to 150 days. Droughts occurs once in every three to five years in these districts. Rainfall is erratic and ill distributed in time and space resulting in frequent crop failures that occur three out of five years. In drier parts of Southern Africa, the coefficient of variation in annual rainfall is between 20 and 40%, while the variation in yield is from less than 15 to 60% with high implications on food security (Lumsden and Schulze, 2007). Land use in the districts is typical of communal lands with dryland crop production in the rainy season and animal rearing. Farmers are dependent on rainfed crop production. The commonly grown crops are maize, groundnuts, sorghum, pearlmillets and groundnuts.

To attest farmers' claim that climate is changing, historical rainfall data from 1980 to 2008 was obtained from Thornhill airbase in Gweru which is 20 km from Lower Gweru while that for Lupane was obtained from Lupani meteorological office. This data was used to determine starting dates, ending date and maximum length of dry spells within rainy seasons.

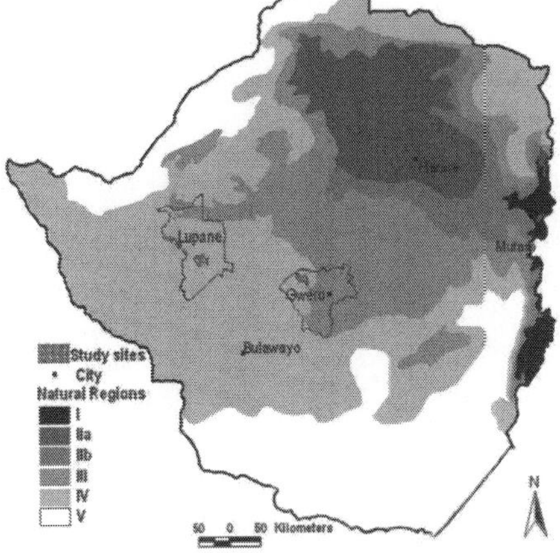

Figure 1: Map of Zimbabwe showing natural regions and locations of the study sites.

The seasonal climate forecasts of the study areas were obtained from the Zimbabwean Met department through the Southern African Regional Climate Outlook Forum (SARCOF) which coordinates the production of the SCF. The seasonal climate forecast is a product of pre-season forums that bring together climate experts involved in seasonal forecast development relevant to Southern African Development community (SADC) to develop the best forecast for the region through a consensus. The product that comes out of the SARCOFs is regional in nature covering the whole of the SADC. After the SARCOF, the national meteorological services, for example the Zimbabwean Meteorological Department downscales the regional product to their country-specific rainfall forecasts.

Participatory Research with Farmers

Three participatory workshops were held with farmers in each of the two districts to get an understanding of the farmers' perceptions of changing climate conditions, to document the indicators they use to come up with their forecast indigenous seasonal climate forecast (ISCF) and also to evaluate field experimental results based on decisions from their indigenous knowledge forecasts. The participating farmers came from two districts (Lower Gweru and Lupane), two wards per district and three villages per ward. Five farmers from each village were selected randomly making a total of 60 farmers who participated in the research. For their respective villages, farmers were asked to come up with prediction for the 2008/2009 and 2009/2010 seasons forecast using their local indicators and their perceptions of changing climate conditions before they went into plenary to come up with a consensus of the seasonal climate forecast for the district. The indigenous seasonal climate forecast was compared with the seasonal climate forecast issued by the Department of Meteorological Services.

Farmers were then asked what management options they would take, given the seasonal climate forecast. In a participatory manner and together with the farmers, field experiments were designed with the aim of testing how farmers' variety selection, tillage method, and fertilizer type and amount decisions, based on the seasonal climate forecast, affect crop yields. Two experiments were planted out in each district for the two consecutive seasons. The 2008/2009 experiment had a split-split plot treatment structure arranged in a Randomised

Complete Block Design (RCBD). Tillage was the main plot factor while variety was the sub-plot factor and fertilizer sub-sub plot factor.

The 2009/2010 experiment had a 2 x 2 x 2 x 2 factorial treatment structure arranged in a RCBD. For each treatment there were two tillage systems, two varieties, two fertilizer levels and two weeding times. The treatments were replicated three times.

RESULTS AND DISCUSSION

Farmers' Perceptions of Climate Change

Farmers indicated that climate has been changing in the previous five years, in that rains start earlier, rains end latter and the maximum number of dry spells has increased. Comparison of the meteorological records with farmers' assessment of climate change showed a large disparity, with few of the stated changes being evident in the long term record. For example, farmers claimed that the rains were starting later and finishing earlier. Analysis of the rainfall record showed that starting rains (20 mm in 2 days) have been later by 5 to 10 days in the last five years compared to the 1980-2008 period at Lupane site but, for Lower Gweru District, it actually started earlier, by 9 days (Table 1). There was no evidence at any one site that the rain season was finishing earlier. Data on dry spells shows that there was an increase in the number of dry spells in the rainy season in both districts. This suggests that farmers might have poor memories of the rainfall variability over the longer term, and given their almost non-existent measurement of rainfall thereby highlighting the need to cross-check (with measured data) farmer derived information about perceived climate change. A similar result was obtained by Moyo et al. (2013) who observed that climatic data show no evidence that corroborates the farmers' perceptions.

Table 1: Rainfall starting dates, ending dates and maximum length of dry spells in the last 5 years and between 1980 and 2008 in Lower Gweru and Lupane Districts

District		Lower Gweru			Lupane	
(Median dates)	Rainfall onset	Rainfall ending dates	Maximum length of dry spells within rainy season	Rainfall onset	Rainfall ending dates	Maximum length of dry spells within rainy season
1980-2008 Last 5 years	Nov 14 Nov 5	Feb 22 Mar 1	16 20	Nov 12 Nov 22	Feb 9 Mar 25	18 20

For dry spells: maximum length of dry spells (Dec 1 to Mar 31).

Table 2: Shows the indicators that farmers use to predict the rainfall season

Wet year	Drought year
(i) Rhus Lancea and Lannea discolor trees produces lots of fruits	(i) Rhus Lancea trees produces few fruits
(ii) Azanza garkeana do not fruit well	(ii) Lannea discolor produces fruits but aborts them
(iii) Heat wave experienced	(iii) Before the rains
(iv) Early haziness soon after winter	(iv) Extended winter period
(v) North easterly winds	(v) North easterly winds dominant
(vi) Frogs turning brownish	(vi) White frogs appear in trees
(vii) Water birds making a lot of noise	(vii) Lots of thunderstorm without rains
(viii) Butterflies seen hovering in the air from north to south starting in October	viii) Early rains starting from early October

Comparison of Scientific and Indigenous Forecasts with Measured and Long Term Rainfall Records

Farmers use a variety of indigenous knowledge systems in predicting weather. Table 2 shows the various indicators used to predict whether the coming season is dry or wet in the two districts. The indicators that the farmers are using are similar to those that have been recorded by other researchers elsewhere in the country, Zambia and Malawi (Kalanda-Joshua et al., 2011; Chagonda et al., 2010; Mubaya, 2010).

The scientific seasonal forecast for Lower Gweru was Normal to Above Normal for the period October, November and December (OND) while normal rains were predicted for the period January, February and March (JFM) in the 2008/2009 season (Table 3). The forecast for Lupane was Normal to below normal for the period OND while normal to below normal rains were expected for the period JFM in the 2009/2010 season. In the 2009/2010 season, the forecasts were below normal to normal in the two districts for both periods OND and JFM.

There was impressive consistency between farmer predictions on seasonal rainfall using indigenous knowledge of environmental indicators and the actual rainfall totals. The indigenous knowledge system predicted a wetter 2008/2009 and drier 2009/2010 season while actual rainfall total was 944 mm for 2008/2009 and 562 mm for 2009/2010 season respectively. The SCF in 2009/2010 predicted normal to below normal rainfall, while farmers at Lower Gweru and Lupane also forecast a poor rainfall season using local indicators but actual rainfall totals were 505 and 617 mm at Lower Gweru and Lupane respectively for the two seasons. It is likely that the deviation of the observed and predicted rainfall could be on one hand due to the fact that the scientific forecasts could not be downscaled enough to capture variation over a small area which could be too coarse to address local farmers' needs (Mahoo and Mpeta, 2010).

Table 3: Comparison of scientific and indigenous forecasts with measured and long term rainfall records

	2008/2009		2009/2010	
Variables	OND	JFM	OND	JFM
L Gweru	Above Normal	Normal	Below Normal to Normal to	Below Normal to Normal
Lupane	Normal to Below Normal	Normal to Below Normal	Below Normal to Normal	Below Normal to Norma
Farmer				
Prediction for both Lower Gweru and Lupane	Wet		Dry	
Actual rainfall	784 mm (Lower Gweru) 1105mm (Lupane)		505 mm (L Gweru) 617 mm (Lupane)	
Long term rainfall range	Lower Gweru Lupane	650 – 800 mm Lupane 450 – 650 mm		

Table 4: Treatments for the trials in the two districts during the 2008/2009 and 2009/2010 seasons

Variables	2008/2009	2009/2010
Tillage	CP	CP
	Ridge	Ridge
Cultivar	SC403, SC513, OPV	SC403, SC513
Fertility	0, 4.2 tons manure,31 kg/ha N, 66 kg/ha N	31 kg/ha N, 66 kg/haN
Weeding	non	1 weeding
		2 weeding

CP- conventional ploughing, SC403 Very early maturing maize variety (110-120 days), SC513 early to medium maturing maize variety (120-130 days). OPV: medium to long season variety (125- 140 days).

Scientific seasonal forecasts generally have greater accuracy for drier seasons that are associated with El Nino events in Southern Africa (van Heerden et al., 2002; Nicholson and Selato, 2000; Lindesay et al., 1986). The 2009/2010 season was El Nino and it was better predicted than the 2008/2009 season by the scientific climate forecast and therefore influenced the prediction of normal to below normal rainfall over Zimbabwe. However, the 2009/2010 season was predicted dry by the indigenous knowledge forecast and the observed rains were below the normal range which is consistent with the scientific forecast's normal to below normal rains.

Response to Seasonal Climate Forecasts

Table 4 shows the management options that farmers in the two districts settled in order to respond to the 2008/2009 and 2009/2010 seasonal climate forecasts. Some of the treatments are different in the two years since the forecasts were different hence different management practices were proposed by the farmers in response to the forecasts therefore inter-seasons results cannot be compared.

The 2008/2009 season was predicted to be above normal in the first half and normal to below normal in the second half of the season. Farmers decided to have fertility treatments so as to try different levels of nitrogen fertilizers and manure. The reasons for the inclusion of the manure treatment was twofold; 1) to demonstrate the usefulness of the material as a fertilizer for crop production, and 2) to demonstrate that it can be a cheaper alternative source of plant nutrients that can cut down the fertilizer cost. Farmers recognized that soil water management to retain most of the rainfall was very important so as to maximize water use efficiency they therefore decided to have tied ridges (Table 4). Since the season was predicted to be normal to above normal in the first half and normal to below normal in the second half of the season, a short season variety was expected to do well hence "SC403" was chosen and compared with a long to medium season variety. An open pollinated variety was chosen because farmers had poor access to hybrid seed, and because they wanted to compare its performance with the hybrids in a season that was not that good.

Table 5: OND and JFM rainfall for the 2008/209, 2009/2010 and historical records (1950-2005) for Lupane and Lower Gweru respectively

Variables	2008/ 2009	2009/ 2010	Long term	2008/ 2009	2009/ 2010	Long term	Long term totals
	OND	OND	OND	JFM	JFM	JFM	
Lupane	331	231	208-258	453	351	333-385	450-650
Lower Gweru	445	222	201-300	403	280	301-401	650-800

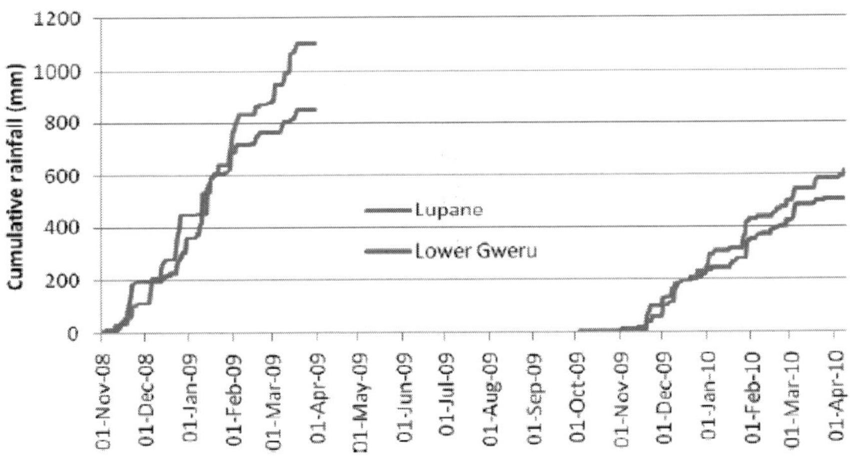

Figure 2: Cumulative rainfall in Lupane and Lower Gweru District.

The 2009/2010 season was predicted to be below normal to normal in both OND and JFM. Farmers came up with varying strategies that ranges from growing of short season cultivars, small grains, water harvesting techniques, planting with the first rains, deep ploughing, stagger planting, application of lower than recommended fertilizers, thorough weed management among others. Low fertilizer levels were chosen since low rains were expected while weeding was very important to reduce moisture competition between crops and weeds. They maintained short season varieties (SC403) that mature early and compared it with a medium season variety (SC513). Tied ridging to capture *in-situ* rainfall was also recommended during this season.

Farmers came up with different management practices for the two years that were contrasting in climate. For instance the use of a short season variety irregardless of forecast outcome, would ensure that they at least are assured of some harvest. This shows that they are risk averse and would want to minimize losses. Sivakumar (2006) observed that farmers have to take a number of crucial crop management decisions and land and water management decisions before and during the growing season, based on climate information. Hansen (2005) and Huda et al. (1991) demonstrate that the ability to understand, monitor and predict weather provides an opportunity to evaluate alternate management strategies for decision-making that takes advantage of good years whilst minimizing losses during unfavorable years.

Field Experimentation

Recorded Rainfall

Figure 2 shows the cumulative rainfall for the two study sites. The rainfall totals for 2008/2009 were 1105 and 784 mm for Lupane and Lower Gweru respectively. The OND and JFM rainfall totals were above the long term range for both Lupane and Lower Gweru during the 2008/2009 season while they were within the long term range during the 2009/2010 season in Lower Gweru only (Table 5). In 2009/2010 total rainfall was 617 and 503 mm respectively for Lupane and Lower Gweru Districts.

Table 6: Effect of fertiliser levels on maize yields during the 2008/9 and 2009/10 seasons in Zimbabwe

Treatment		2008/2009			2009/2010	
	Daluka	NyamaMudubiwa		Daluka	NyamaMudubiwa	
67 kg/ha N	4.063.	1.603	3.823	1.446a	1.9003	0.3313
31 kg/ha N	3.3436	0.8736	2.786	0.9756	1.2426	0.164b

Manure	2.760b	0.641b`	1.50`			
Zero N	2.140b	0.414`	1.13`			
*LSD	0.898	0.319	0.595	0.297	0.390	0.101

*Number with the same letter are not significantly different, LSD: Least Significant Difference.

Table 7: Effect of weeding times on maize yields during the 2009/2010 season in Zimbabwe

Treatment	Daluka	Nyama	Mudubiwa
	2009/2010	2009/2010	2009/2010
Weed 1	1.0333	1.5673	0.1933
Weed 2	1.388″	1.5833	0.252″
LSD	0.297	0.390	0.101

*Number with the same letter are not significantly different.

Table 8: Effect of varieties on maize yields during the 2008/2009 and 2009/2010 seasons in Zimbabwe

		2008/ 2009			2009/ 2010	
Treatment						
	Daluka	Nyama	Mudubiwa	Daluka	Nyama	Mudubiwa
SC403	3.43	0.9053	2.543	1.233	1.85e	0.2533
SC510	2.733	0.8803	2.163	1.23	1.283b	0.2423
OPV	3.13	0.8613	2.223			
LSD	0.660	0.276	0.515	0.297	0.390	0.101

Experimental Results

Adding recommended amounts of nitrogen (67 kgN/ha) gave significantly higher yields than adding low fertilizers levels, manure and no nitrogen at all in the sites except during the favorable 2008/2009 season when between 800 and 1100 mm of rainfall was received in Daluka ward (Table 6). No significant differences in yield were recorded between adding recommended amounts of nitrogen and adding low amounts of nitrogen (31 kgN/ha) during this season. 2008/2009 was a wetter year and we would have expected yield differences between adding 67 and 31 kgN/ha of nitrogen. Adding 67 kgN/ha of fertilizers gave significantly higher yields than adding 31 kgN/ha of nitrogen at all the three sites during the 2009/2010 season.

Weeding two times gave significantly higher yields than weeding once at Daluka and Mudubiwa while there were no significant differences at Nyama (Table 7). It has been observed that weed competition results in crop yield losses hence weed control improves maize yield grain (Abouziena et al., 2008; Dalley et al., 2006). There were no yield differences attributed to variety in both seasons and at all the three sites (Tables 7 and 8) during the 2008/2009 season and the 2009/2010 season accept at Nyama where the short season variety (SC403) outperformed the medium season variety (SC513).

The treatments that were recommended by farmers for adapting for below average season did not perform better than those that were meant for an above average season. For example, the recommended fertilizers levels resulted in better yields than low fertilizer levels that are meant for below average rainfall predicted in 2008/2009 season but ended up having higher than expected rainfall amounts. Studies elsewhere have demonstrated that there are no yield gains emanating from adding high amounts of fertilizers in drier seasons and differences are only obtained in wetter seasons (Mburu et al., 2011).

However, the short season varieties that are meant for below average season outperformed the long season varieties possibly because of the short length of the growing season. The suggested tied ridges and ripping that are meant to increase soil moisture under drier conditions did not give yield differences because the tied ridges were destroyed by heavy downpours during the early part of the seasons.

CONCLUSIONS

Farmers can only forecast a wet or dry season from their indicators while the scientific SCF indicate whether is normal, below normal or above normal for two periods of the rainy season, OND and JFM. Results from this study demonstrate that, in the absence of the scientific SCF, farmers can use ISCF which has proved to be consistent with each other in the two years. It is also important that SSCF and ISCF be integrated thereby imbedding adaptation strategies in communities' existing knowledge of climate variability and indigenous prediction systems which is being recommended by Huq and Reid (2007). The study demonstrates that seasonal climate forecast is an ideal entry point for bringing meteorological officials, researchers and extension agents together in working with the farmers in coming up with crop management practices that respond to the seasonal climate forecast thereby increasing their crop yields. From discussions held with the farmers and the nature of experiments proposed by the farmers, in response to the 2008/2009 and 2009/2010 forecasts (both modern and indigenous knowledge), it was evident that farming decisions are influenced by the seasonal climate forecast which include decisions on the following: choice of crop and variety, tillage systems to use, planting date, fertilizer amounts to use and when and how to weed.

ACKNOWLEDGEMENTS

This publication is an output from a project funded by CCAA/IDRC and DFiD (Grant number 104144). The views expressed are not necessarily those of CCAA/IDRC/DFiD. The authors would particularly like to thank the communities who participated in the research for their assistance.

REFERENCES

1. Abouziena HF, El-Metwally IM, El-Desoki ER (2008). Effect of plant spacing and weed control treatments on maize and associated weeds in sandy soils. Am. Eur. J. Agric. Environ. Sci. 4(1):9-17.

2. Bharara LP, Seeland K (1994). Indigenous knowledge and drought in the Arid zones of Rajasthan: Weather prediction as a measure to cope with a hazardous climate. Int. Asienforum 25(1/2):53-71.

3. Chagonda I, Churi J, Dieye A, Houenou B, Hounkponou S, Kisiangani E, Kituyi E, Lukorito C, Macharia A, Mahoo H, Majule A, Mapfumo P, Mtambangwe F, Mugabe FT, Ogallo L, Ouma G, Wanda G (2010), Integrating meteorological and indigenous knowledge-based seasonal climate forecast for the agricultural sector: Lessons from participatory action research in sub-Saharan Africa. CCAA learning paper. In: Ziervogel, Gina; Opere, Alfred (eds.) P. 19.

4. Challinor AJ (2008). Towards a science of adaptation that priorities the poor. Poverty in a changing climate, TANNER, T AND MITCHELL, T (eds). IDS Bull. 39(4):81-86.

5. Dalley CD, Bernards ML, Kells JJ (2006). Effect of weed removal timing and row spacing on soil moisture in corn (*Zea mays*). Weed Technol. 20:399-409.

6. Hansen JW (2005). Integrating seasonal climate prediction and agricultural models for insights into agricultural practice," Philosophical transactions of the royal society. *Biological sciences:* Royal Society Publishing 360(1463):2037-2047.

7. Huq S, Reid H (2007). Community-based adaptation An IIED Briefing: A vital approach to the threats climate change poses to the poor, IIED Briefing, London: International Institute for Environmental Development.

8. IPCC (2001). Climate change 2001: Impacts, adaptation and vulnerability. Technical Summary, IPCC Publication.

9. IPCC (2007). Climate Change 2007: The Physical Science Basis. Contribution of Working Group 1 to the Fourth Assessment Report of the Intergovernmental Panel on Climate Change, Chapter 11. Regional Climate.

10. Kalanda-Joshua M, Ngongondo C, Chipeta L, Mpembeka F (2011). Integrating indigenous knowledge with conventional science: Enhancing localised climate and weather forecasts in Nessa, Mulanje, Malawi. J. Phys. Chem. Earth doi:10.1016/j.pce.2011.08.001

11. Lasage R, Aerts J, Mutiso GCM, de Vries A (2008). Potential for community based adaptation to droughts: sand dams in Kitui, Kenya. Phys. Chem. Earth 33:67–73.
12. Lindesay JA, Harrison MSJ, Haffner MP (1986). The Southern Oscillation and South African rainfall. S. Afr. J. Sci. 82:196–198.
13. Lumsden TG, Schulze RE (2007). Application of Seasonal Climate Forecasts to Predict Regional Scale Crop Yields in South Africa, in Climate Prediction and Agriculture: Advances and Challenges. Hansen, J.W. and Sivakumar, M.V.K. (EDS). Berlin: Springer pp. 213-224.
14. Mahoo H, Mpeta M (2010). Combining indigenous and scientific weather forecast knowledge in climate risk management in semi-arid areas of Tanzania. Paper presented at the workshop on "Achieving benefits of enhanced service delivery by national meteorological services in eastern and southern Africa" Dar es Salaam.
15. Mburu DM, Lenga FK, Mburu MWK (2011). Assessment of maize yield response to nitrogen fertiliser in two semi-arid areas of Kenya with similar rainfall pattern. Jomo Kenyatta Univ. Agric. Technol. 13(1): 22-34.
16. McCrea R, Dalgleish L, Coventry W (2005). Encouraging use of seasonal climate forecasts by farmers. Int. J. Climatol. 25:1127-1137.
17. Meinke H, Hochman Z (2000). Using seasonal climate forecasts to manage dryland crops in northern Australia: experiences from the 1997–98 seasons. In Applications of Seasonal Climate Forecasting in Agricultural and Natural Ecosystems: The Australian Experience, Hammer C.L., Nicholls, N., Mitchell, C. (eds). Kluwer Academic: The Netherlands pp. 149–165.
18. Moyo MM, Kunzekweguta BM, Mazvimavi K, Craufurd P, Dorward P (2012). Farmer perceptions on climate change and variability in semiarid zimbabwe in relation to climatology evidence. Afr. Crop Sci. J. 20:317-335.
19. Mubaya CP (2010). Farmer strategies towards climate variability and change in Zimbabwe and Zambia. DPhil. Dissertation, University of the Free State. Centre for Development Support.
20. Nicholson SE, Selato JC (2000). The influence of La Nina on African rainfall, Int. J. Climatol. 20:1761–1776.

21. Nyong AO, Adesina F, Osman EB (2007). The value of indigenous knowledge in climate change mitigation and adaptation strategies in the African Sahel. Mitig. Adapt Strat. Glob. Change 12:787–797.
22. Nyong AO, Kanaroglou PS (1999). Domestic water demand in rural semi-arid north-eastern Nigeria: identification of determinants and implications for policy. Environ. Plan. 34(4):145–158.
23. Parry M, Rosenzweig C, Iglesias A, Fisher G, Livermore M (1999). Climate change and world food security: a new assessment, Global Environ. Change 9:S51-S67.
24. Patt A, Gwata C (2002). Effective seasonal forecast applications: examining constraints for subsistence farmers in Zimbabwe. Global Environ. Change 12:185-195.
25. Patt A, Suarez P, Gwata C (2005). Effects of seasonal climate forecasts and participatory workshops among subsistence farmers in Zimbabwe. Proceed. National Acad. 102(35):12623–12628.
26. Prowse M, Scott L (2008). Assets and adaptation. Poverty in a changing climate, Tanner, T and Mitchell, T (eds). IDS Bull. 39(4):42-49.
27. Sivakumar MVK (2006). Climate prediction and agriculture: current status and future challenges. Climate Res. 33:3-7.
28. Van Heerden J, Terblanche TE, Schulze RC (2002). "The Southern Oscillation and South African summer rainfall," J. Climatol. 8:577– 597.
29. Vincent V, Thomas RG (1960). Agricultural Survey of Southern Rhodesia. Part 1. Agro- Ecological Survey. Government Printers, Salisbury.
30. Ziervogel G, Calder R (2003). Climate variability and rural livelihoods: assessing the impact of seasonal climate forecasts. Area 35:403-417.

Chapter 3

Effect of Substrate on Vegetative Growth, Quantitative And Qualitative Production of Muskmelon (cucumis Melo) Conducted in Soilless Culture

Afrâa Radhouani, Mokhtar El Bekkay, and Ali Ferchichi

Dry Lands and Oasian Cropping Laboratory, Institute of Arid Regions, El Jorf 4119, Medenine, Tunisia. Accepted 24 December, 2010

ABSTRACT

Greenhouses heated and irrigated by geothermic water, constitute a promoter sector in the South of Tunisia. However, simultaneous salinity of sand and water of irrigation limits amply, the development of this sector. To control salinity, it seems efficient to adopt interesting agricultural techniques like soilless culture. The selection of suitable media is the key that ensures the success of this technique, seeing that the adoption of not standardized one, limits the correct nutrient solution management so the development of plant. The substrate selection must consider technical and economic implications. In this framework, two ubiquitous substrates of the South of Tunisia, sand and compost of dry palms, were compared to the imported media, perlite, for soilless culture of muskmelon (*Cucumis melo*). This study was based on the vegetative growth, quantitative and qualitative production. Results have shown that local substrates have simulated, in relation to perlite, strong and deep roots that permit an effective absorption of nutrients. Subsequently, leaves were larger, their content in fresh and dry matter were higher and stems were taller. This superiority was sustained by superiority of index of growth, NAR. This behavior has reverberated on precocity and yield that were greater on these two media. As regard to external characteristics of fruits, firmness and diameter were higher particularly on compost. With exception of pH, gustative quality was not affected by type of substrate.

INTRODUCTION

Nowadays, seeing the increase of human population, an emphasis on higher vegetable production all year round is noted. To response to this demand, sustainable development aims towards upcoming the target of optimizing crop yield by intensifying agriculture (Maloupa, 2000). This intensification conjoined with the shortage of water resources and arable lands can be achieved only through the adoption of promising agricultural practices such as soilless culture. Soilless culture covers all methods and systems of production tool using mineral solution for the plant nutrition with another substrate or support than soil (Resh, 1989; Robin, 1998; Butt et al., 2004; Sheikh, 2006; Gruda, 2009) such as water culture, gravel culture, aeroponics, tube culture and nutriculture (Schwarz, 1994). It can be done on open or closed system (Baas et al., 1995; Papadopoulos et al., 1999)

It was advisable to gradually extend this agricultural practice due to its advantages (Jiang and Yu, 2004; Metin-Sezen et al., 2006; Yetisir et al., 2006). Indeed, it is an efficient tool to overcome problems associated with production factors (Verdonck, 1975; Granqvist, 1981;

Olympios, 1992; Abd El-Hady and El-Dardiry, 2006) and create possibility of cultivation with flexibility even in regions where natural growing conditions are hostile (Schwarz, 1994; Morard, 1995; Grillas et al., 2001). The evidence that, this technique helps to surmount soilborne diseases and pests seeing that, the distribution of roots pathogens is drastically minimized (Schwarz, 1994; Choukr-Allah, 2004) has widespread, too, rapidly this practice.

Table 1: Characteristics of local sand

	Gravel	2.7
	Rough sand	46.3
Granulometric composition (%)	Fine sand	48.3
	Silt and clay	2.7
pH		8.2
EC (mS/cm)		2.5
Rate of retention of water (%)		26

Incontestably, this system of culture ensures an economy of water in relation to conventional soil culture due to decrease of soil evaporation and seepage (Schwarz, 1994; Schröder and Lieth, 2002) and endows an accurate control of the nutritional supply of plants (Resh, 1978; Schwarz, 1994; HO, 2004; Sheikh, 2006; Gruda, 2009). Therefore, this technique allows the use of brackish water (Adams, 1991) namely with the scarcity of good quality of water because of the adjustment of nutrient solution and its permanent replacement. Indeed, improving water use efficiency and the irrigation with available saline water in an economically viable manner, are the main objectives of introducing such technique (Qaryouti et al., 2006). Likewise, among the alternatives to soil sterilization by methyl bromide, soilless culture represents one of the more widely considered possibilities (Ozeker et al., 1999; Maloupa, 2002; Savvas, 2003; Lieten et al., 2004; Giménez et al., 2008; Kaya et al., 2008; Leoni and Ledda, 2004). As well, it protects soil and underground water from pollution caused by pesticides and fertilizers particularly nitrogen (Yetisir et al., 2006). The choice of substrates for soilless cultivation, represents one of the key points to be considered (Leonardi, 2006) due to technical and economical implications (Giuffrida et al., 2008). Indeed, the choice of a not standardised growing media is no guarantee for a correct nutrient solution management. There is no univocal scheme for the choice of the growing media (Leonardi, 2006). But, the knowledge of its characteristics is very important for this selection, seeing that the properties of different materials used as growing substrates, exhibit direct and indirect effects on plant growth and production (Verdonck et al., 1981).

These characteristics concern physical, chemical and biological properties. Physical properties concern aeration, drainage and water retention capacity (Blanc, 1987; Lemaire et al., 1989; Cabrera, 2003). Chemical and biological features are indicated, respectively, by not having toxic material and very limited and/or less pathogen and pest (Blanc, 1987).

Soilless culture is in the process of becoming an interest part of agriculture in Tunisia and gained crescent attention among farmers especially under greenhouses heated and irrigated by geothermic waters (Haddad, 2007). The soilless culture systems which are currently used in Tunisia, are slanting sand bed and pudding bags. Tunisia is in root of vulgarizing enormously, the use of this practice under these greenhouses, in order to overcome the well-known problems of salinity and pathogenic agents that invade this sector (Kouki, 1999). However, it remains limited for big geothermic societies because of the high installation cost compared to soil (material disposal and raising), the use of sophisticated control methods and its requirement of high technical level (Radhouani et al., 2008). The major growing media are imported with a high use of perlite. Nevertheless, several studies have attested that, this growing medium constitutes the most promising and efficient substrate in soilless culture (Grudina et al., 1994; Park et al., 1999), it seems to be of great interest to search locally available and less costly substrates (Verdonck et al., 1981; Klougart, 1983; Abd El-Hady and El-Dardiry, 2006; Abd El-Hady et al., 2006; Giménez et al., 2008; Tzortzakis and Economakis, 2008; Abo-Rezq et al., 2009; Abd ElHady and Shanan, 2010). To investigate this issue, a trial of a culture of muskmelon conducted under greenhouse, heated and irrigated by geothermic water was carried out. This study aims to test the effects of two local substrates, sand and compost, and imported one, perlite on this plant.

MATERIALS AND METHODS

Location, Plant Material, and Growth Conditions

The study was carried out in a 255 m^2 mono tunnel (8.5 x 30 m) covered with a 200 µm polyethylene film, at the experimental field of Institute of Arid Regions (IRA). This greenhouse was heated by geothermic waters. Heating was realized by the circulation of geothermic waters (60°C) in corrugated polypropylene pipes (PPØ 25) placed on the plastic between plant rows. Control of daily temperature occurred by lateral aeration. Transplant took place in early January 2007, at a density of 3 plants/m^2, using the hybrid F1 'Calypso', considered as one of the most representative muskmelon varieties cultivated in Tunisia; seeing its high crop performance and fruit quality. Plants were grown under natural light conditions. They were grown with two stems and pruned above the fifth cluster. An open soilless cultivation system was adopted. Three grown media were considered:

- Perlite: Perlite used in this study had 90% volume of its expanded particles in 5 to15 mm range, and a very high air-filled porosity of around 60%;
- Local sand with characteristics represented in Table 1;
- Local compost formed by fermentation of dry palms with addition of manure (Haddad, 2003; Bouhaouach et al., 2009). The main characteristics of this media are summarized in Table 2.

The adopted system is pudding bags culture system: plastic bags with volume of 33 L and U shape were used as containers with density of three plants per container. They were placed on ground, settled down beforehand and covered by a plastic film. They were disposed on a fine layer of graves. They were perforated to drain excess of water.

Table 2: Characteristics of local compost

C/N (%)	Organic matter (%)	Total porosity (%)	pH	EC (ms/cm)	Rate of retention of water (%)

27.1	60	62.2	7.64	4.1	31

Table 3: Chemical composition of nutritive solution (%)

	N	K	Ca	Mg	H2PO4 -	SO4 2-
Water of irrigation	0	1.06	12	7.2	0	14
Norms of fertilization	14.3	6	8.5	2	1.3	2.5
Nutritive solution	14.3	4.94	0	0	1.3	0

Table 4: Composition of nutritive solution regards to stage of development of plants (Huguet et al., 1971)

Stage of growth chemical	Plantation-first fruit set	First fruit set - last fruit set	Last fruit set- start of harvest	Start of harvest- end of harvest
NH4NO3	20	100	71.5	94
KNO3	19	100	73.8	45
H2PO4	17	88.5	80	100
HNO3	20	100	71.5	94

Nutrient Solution Management

The nutritive solution was pumped using a pump with 1 atm power,through an open drip irrigation system with one emitter per plant and a flow rate of 4 L/h. This solution was formulated according to the chemical composition of water of irrigation and norms of fertilization of muskmelon (Table 3). The nutritional needs of plants were determined referring to Huguet et al. (1971) (Table 4). Plants were fertilized by fertigation. The plants were irrigated daily 4 to 5 times, depending on the size of the plant and the growing climatic conditions. Additionally, the volume of irrigation water was adjusted to ensure a leaching fraction, sufficient to contain the increase of salt in the substrate (Kempkes and Stanghellini, 2003). To avoid this increase, plants were over watered with a drainage volume equal to about 80% of the supply once a week, with geothermic waters without nutrients. The experiment plan was of a completely randomized design with three replications and each one was represented by 24 plants per row (three rows in the greenhouse).

Data Collection and Analysis

The experiment was conducted as a randomized complete block design with three replicates. Each experimental unit consisted of 24 plants per line. Evaluation of the effect of substrate on vegetative growth has included steam (average height) and leaves (fresh matter, rate of dry matter and relative content of water). The measures were recorded on the fifth leaf from the top, seeing that it corresponds to a transformation from the state of well to source, as it was indicated by Ouled et al. (2006). Dry matter and leaves' area have served to determine growth indexes: RGR (Relative Rate of Growth) and NAR (Net Assimilation Rate). These measures were done 49, 64 and 79 days after transplantation. At the end of culture, plants were uprooted and length, fresh and dry weight of roots were measured. Besides, volumes of roots expressed as volume registered on the base of water moved by roots were specified. For each substrate, days preceding harvest were counted. The average number and weight of fruits were determined. Fruits were classed into three classes according to their weights (CTIFL, 1991): C1: weight inferior to 600 g; C2: weight ranging between 600 and 900 g; C3: weight superior to 900 g. Total yield per substrate was determined. Firmness of fruits was estimated by measuring their penetrance. The incidence of substrate on gustative quality was estimated by measuring electrical conductivity, EC, using a conductive meter, pH (by an electronic pH) and soluble solids determined by a refractometer and expressed in °Brix, of juice's fruits. Its titrable acidity was measured by titration of 25 ml of homogenated juice with 0.1 NaOH until pH 8 and reported in percentage of acid content. Statistical analyses were performed using a level of 0.05 (5%) for the ANOVA and Tukey's post hoc tests. Differences between the means of the three substrates were compared, using the least significant difference (LSD). Levels of significance are represented by *(P < 0.05), ** (P < 0.01) and *** (P < 0.001) and NS (not significant).

RESULTS AND DISCUSSION

Effect of Substrate on Vegetative Growth

Roots of plants developed on sand and compost have shown a tremendous superiority through their volume and fresh weight ($P < 0.001$) in comparison to those of plants cultivated on perlite (Table 5), due to the enormous development of lateral thin roots (Figure 1). With the local substrates, plants height was statistically similar and significantly greater than the one recorded on perlite: with reference to the latter, plants have reached as at 79 days after transplantation, an average length of stem of around 125.00 cm; yet, this parameter was about 185.22 and 198.88 cm respectively on sand and compost (Figure 2). This superiority was noted, too, with growth of leaves. Indeed, local substrates have presented similarly, fifth leaves with a fresh matter of 1.45 g and an average area of 41.65 cm^2. However, on perlite, growth of leaves was limited to a fresh matter of 1.01 g and an expansion of 31.83 cm^2. Rate of dry matter of leaves for the three treatments were statistically similar (Table 5).

Table 5: Effect of substrates on vegetative growth, quantitative and qualitative production

	Pe rlite	Sand	Compost	Level of significance
Length (cm)	27 a± 1.13	28.33 a ± 1.42	32.66 a±1.72	NS
Volume (ml)	27.33 b ± 0.65	41.66 a ± 0.65	49.66 a ± 1.17	**
Fresh weight (g)	31.23 b ± 0.77	61 a ± 0.79	62 a ± 0.92	***

Dry weight (g)	2.7 b ± 0.68	5.03 a ± 1.07	4.8 a ± 0.67	**
Surface (cm2)	31.83 b ± 0.54	40.5 a ± 0.61	41.83 a ± 0.85	
Fresh matter (g)	1.01 b ± 0.07	1.33 a ± 0.11	1.58 a ± 0.17	
Rate of dry matter (%)	14.51 a ± 0.04	16.29 a ± 0.04	16.68 a ± 0.08	NS
Content of water (%)	83.89 a ± 0.03	83.39 a ± 0.02	81.51 a ± 0.04	NS
Days before harvest	91	90	88	
Fruits/plant	5.33 b	6.32 a	6.28 a	
Kg/fruit	0.41b	0.51 ab	0.65 a	
Total production (Kg/ plant)	2.21 c	3.24 b	4.11 a	
Penetrance (cm)	0.15 a ± 0.048	0.12 a ±0.044	0.04 b± 0.042	
Diameter (cm)	7.66 b ± 0.61	10.56 a ± 0.48	10.42 a ± 0.14	
pH	6.66 b ±0.02	6.87 a ±0.09	6.98 a ± 0.08	
EC (mS/cm)	7.18 a ± 0.40	7.51 a ± 0.52	6.94 a ± 0.32	NS
IR (°Brix)	10.1 a ± 0.80	10.8 a ± 1.60	11.8 a ± 0.65	NS
Acidity (%)	1.40 a ± 0.02	1.19 a±0.14	1.17 a ± 0.13	NS

Values followed by the same letters within each line are statistically similar according to test of Newman and Keuls at P_ 0.05. Levels of significance are represented by * P < 0.05, ** P < 0.01, *** P < 0.001 and NS, not significant

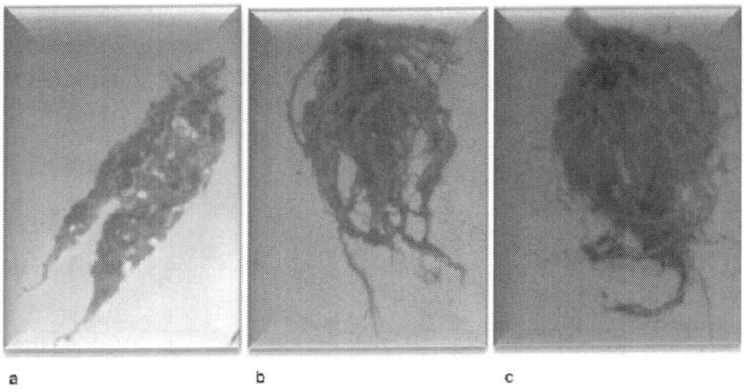

Figure 1: Roots on perlite (a) sand (b) and compost (c).

Although their similar rate of growth, RGR, (Figure 3) substrates have shown a conspicuous difference concerning their net assimilation rate (NAR). For the last parameter, values were negative on three substrates, yet this negativity was more drastic on perlite (Figure 4)

Effect of Substrate on Quantitative Production

Harvest of fruits had begun on compost, 88 days after transplantation with superiority of 2 and 3 days respectively, in relation to sand and perlite (Table 5.

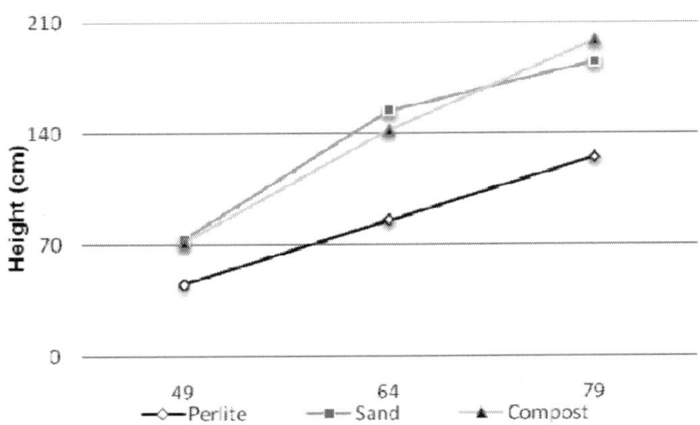

Figure 2: Effect of substrates on the rate of growth (cm/day).

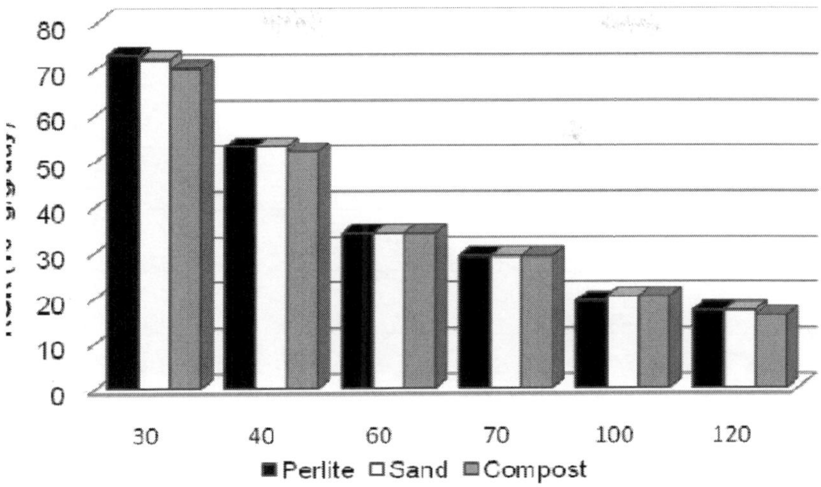

Figure 3: Effect of substrates on relative rate of growth, RGR, (10^{-4} g/g/day).

Plants cultivated on sand and compost had formed towards 6 fruits with an average weight of 0.58 Kg. This effect of two ubiquitous media was higher than one or of perlite, where each plant had formed 5.3 fruits with a small weight around 0.41 Kg (Table 5). All the same, total production was better on local substrates than on perlite. This superiority was about 66.28%. Additionally, these two growing media have enthused production of fruits with marketable weight. Indeed, on sand and compost, this class was represented respectively by 43.64 and 22.44% of total production, but for plants cultivated on perlite, this class constitutes only % of their all production (Figure 5).

Effect of Substrate on Qualitative Production

The most important, firmness, deducted from penetrance of fruit's pericarp, was remarked on compost (Table 5). Fruits harvested on sand and compost were larger (important diameter) than those collected on perlite. Measures have revealed that, the effect of substrate on gustative quality was similar (Table 6). With exception, pH was higher on sand than on the two other substrates.

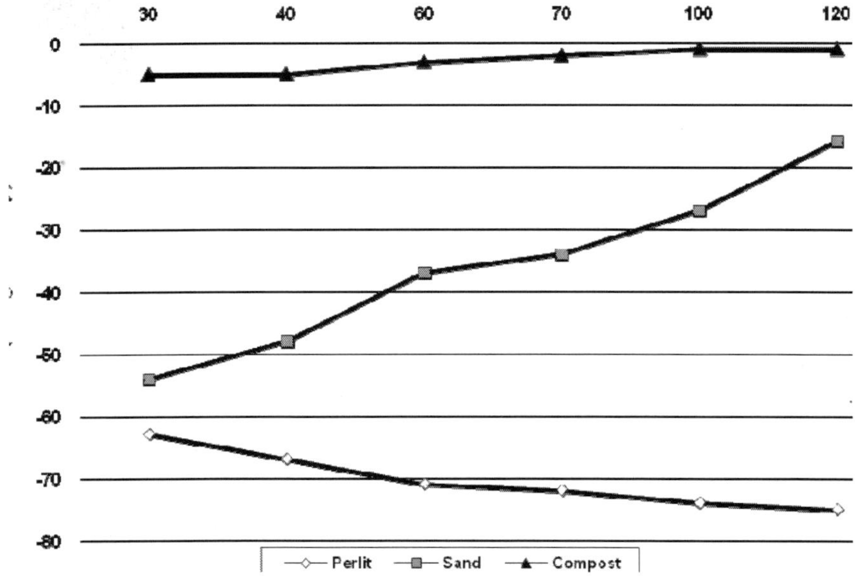

Figure 4: Effect of substrates on net rate of growth, NAR, (10-4 g/cm2/day.

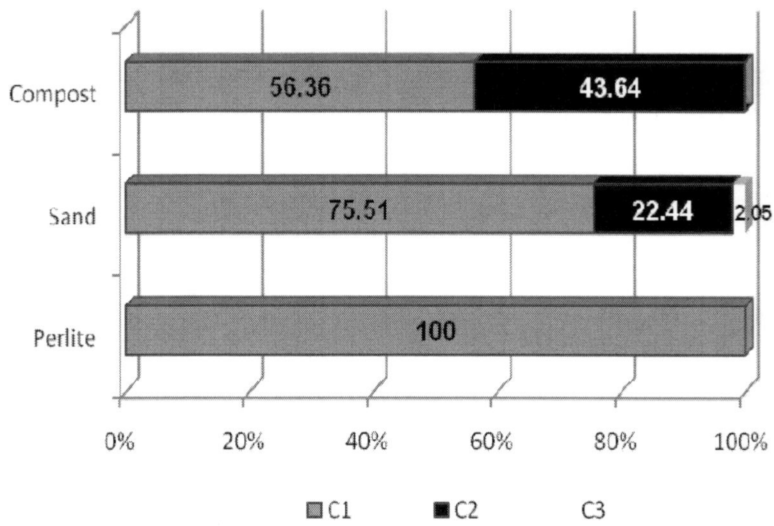

Figure 5: Effect of substrates on caliber of fruits.

DISCUSSION

In the lights of the results of this study, it was seen that local substrates, sand and especially compost can constitute a promising alternative to the use of imported media, such as perlite. Indeed, these substrates have promoted remarkably vegetative growth of plants. This behavior may be attributed to the great growth of roots, represented by their great fresh and dry weights and volume. Moreover, thin roots were dominant on their morphology. The dominance of such roots can be explicated by the high holding water capacity of the two local substrates. Thus, assimilation of nutrients was efficient, seeing that these roots were effective in absorption of brut sap ingredients, as it was affirmed by Daaaloul et al. (2001). Consequently, leaves were larger and richer in fresh matter and stem was taller. The weak effect of perlite, on the height of plant, compared with sand was indicated too for tomatoes by Haddad (2007). The lowest effect of perlite on vegetative growth can be due to the high electrical conductivity of its nutrient solution (Xu et al., 1997). Superiority of growth on sand and compost, in relation to perlite was confirmed by the net assimilation rate, NAR. Indeed, on this medium was recorded the most negativity of this parameter. This negativity may infer that, fifth leaves presented a physiological state of well, as it was indicated by Marcelis et al. (1998) and Loveys et al. (2002). This result confirms the faster vegetative growth of plants cultivated on local substrates, seeing that negative values of this parameter indicates a lengthiness of this growth, as it was attested by Fahrurrozi (2000) for muskmelon. Difference in vegetative growth has reverberated on precocity which was better on local substrates than on perlite, seeing their hasty vegetative growth. Superiority of compost concerning this parameter can be attributed to its heat absorbing ability justified by its dark color, as it was indicated by Yetisir et al. (2006) for basaltic tuff. A part the slight enhancement of precocity by local substrates in relation to perlite, quantitative production was considerably promoted. This finding corroborated the one of Yetisir et al. (2006) elaborated between sand and perlite. It seems that, promising growth of leaves in terms of area and fresh matter was beneficial for production of heavy fruits. Furthermore, fruits with marketable weight were higher on these two substrates. Concerning qualitative production, fruits harvested on local media, especially compost, were firmer and larger than those formed on perlite. However, Haddad et al. (2003) have sustained, for tomatoes, a similarity between these substrates. The superiority of sand in relation to perlite was evidenced, too, by Yetisir et al. (2006) for watermelon. Gustative quality was unaffected by type of substrate. This remark was consistent with previous studies of Haddad et al. (2003) and Haddad (2007) for tomatoes. Yetisir et al. (2006) noted similar total soluble solids for watermelon cultivated on sand and perlite.

CONCLUSIONS

In the South of Tunisia, few years after the use of geothermal waters in protected cultures, decrease of soil fertility and increase in soil salinity, associated to the incurrence of soil born diseases, have rapidly developed and limited amply both the growth and the productivity of crops. To overcome these constraints, a local soilless system culture was developed. This work compared two local substrates, sand and compost, to perlite. Results have shown that, local grown media have revealed promising performance: plants grew faster, produced earlier and formed more fruits with marketable yields. Nevertheless, it was noted at the end of culture that, compost and particularly sand have subside. This disadvantage can be prevailed over by their mixing, to reward their mutual deficiencies.

REFERENCES

1. Abd El-Hady M, El-Dardiry EI (2006). Improving Hydrophysical Properties Quality of Compost. J. Appl. Sci. Res., 2(12): 1137-1141.
2. Abd El-Hady M, Shaaban SM, El-Dardiry EI (2006). Effect of Hydrophysical Improvement of Substrates on Cucumber Plant Growth. J. Appl. Sci. Res., 2(12): 1185-1190.
3. Abd El-Hady M, Shanan N (2010). Enhancement growth characters of Pothos plants (Epipremnum Aureum Lind L.) grown in different improved pot media. Int. J. Acad. Res., 2(2): 89-97.
4. Abo-Rezq H, Albaho M, Thomas B. (2009). Effect of Sand in Growing Media on Selected Plant Species. Eur. J. Sci. Res., 26(4): 618-623.

5. Adams P (1991). Effects of increasing the salinity of the nutrient solution with major nutrients or sodium chloride on the yield, quality and composition of tomatoes grown in rockwool. J. Horticult. Sci., 66: 201-207.

6. Baas R, Nijssen HMC, Van Den Berg TJM, Warmenhoven MG (1995). Yield and quality of carnation (Dianthus caryophyllus L.) and gerbera (Gerbera jamesoniiL.) in a closed nutrient system as affected by sodium chloride. Sci. Hort., 61(1): 273-284.

7. Blanc D (1987). Les cultures hors sols. Ouvrage collectif. INRA. Paris, p.478. Bouahaouach H, Marc HC, Kouki K (2009). Compostage ET valorisation des déchets oasiens pour l'amélioration des sols ET de la productivité. Symposium international AGRUMED: Gestion des Ressources en Eau ET en Sol ET Durabilité des systèmes de Cultures en Zone Méditerranéenne. Rabat, IAV Hassan II. 14-16 Mai 2009.

8. Butt SJ, Varis S, Al-Haq MI (2004). Improvement of sensory attributes of tomatoes (Lycopersicon esculentum Mill) through hydroponics. Agric. Biol., 6(2): 388-392.

9. Cabrera RI (2003). Fundamentals of container media management. Part 1. Physical properties, The State Univ. of New Jersy Agric. Exp. Station.

10. Choukr-Allah R. (2004). Protected cultures in Marocco: New trends and developments. international work shop: "The production in the greenhouse after era of the methyl bromide" Comiso, Italy. 1-3 Apr 2004. CTIFL. 1991. Le melon, p. 295.

11. Daaloul A, Bchini H, Sayar R. (2001). Variabilité génétique de quelques paramètres du système racinaire du blé dur (Triticum durum) sous deux **régimes** hydriques. Biodivers. int., 129: 25-31.

12. Fahrurrozi A (|2000). Effect of mulch / mini-tunnel and thermal-water tube combination on daily dioxide concentration and early growth of musk melons. Thèse de doctorat. Department of plant science. Macdonald. Campus of Mc Gill University. Monterall, Quebec, Canada, p. 138.

13. Giménez G, Andriolo JL, Janisch D, Rodrigo G (2008). Closed soilless growing system for producing strawberry bare root transplants and runner tips. Pesq. agropec. Bras. 43(12): 1757-1761.

14. Giuffrida F, Leonardi C, Marfà O. 2008. Substrate reuse in tomato soilless cultivation. Acta Hort., p. 801, ISHS. 2008. Granqvist G. (1981). Recent experiences in the use of substrates for vegetables production under glass in Sweden. Acta Hortic., 126: 262- 264.
15. Grillas S, Lucas M, Bardopoulou E, Sarafopoulos S (2001). Perlite based soilless culture systems: Current commercial applications and prospects. Acta Hort., 548: 105-113.
16. Gruda N (2009). Do soilless culture systems have an influence on product quality of vegetables? J. Appl. Bot. Food Quality, 82: 141-147.
17. Grudina R, Fadda N, Chessa F, Leoni S, Serra G, Tononi F, Leoni S. (1994). Two years of watermelon cultivation in soilless culture. Act Horticult., 361: 574-577.
18. Haddad M. (2003). Exploitation des eaux géothermiques du Sud tunisien pour la culture de la tomate (Lycopersicon esculentum. Mill). Thèse de doctorat de biologie, p. 132
19. Haddad M (2007). Effect of three substrates on growth, yield and qualitytomato by the use of geothermal water in the south of Tunisia. J. Food Agric. Environ., 5(2): 175-178.
20. Haddad M, Boukhriss M, Mougou A (2003). Growth and quality of tomato plants grown in sand and perlite by use of geothermal water. International Symposium on Managing Greenhouse Crops in Saline Environment. Pisa. Italy. 9-12 July 2003.
21. HO LC (2004). The contribution of plant physiology in glasshouse tomato soilless culture. Acta Hort., 648: 19-25. HUGUET.
22. Jiang WJ, Yu HJ (2004). Current situation and perspectives of protected horticulture in china. International work shop:" The production in the greenhouse after era of the methyl bromide". Comiso, Italy1-3 Apr. 2004.
23. Kaya S, Caturano E, Tuzel Y, Leonardi C (2008). Response of tomato plants to organic nutrition in soilless culture. J. Food Agricult. Environ., 6(2): 303
24. Kempkes F, Stanghellini C (2003). Modelling salt accumulation in a closed system: A tool for management with irrigation water of poor quality. Acta Hort., 641: 143-148.
25. Kouki K (1999). Protected cultivation in Tunisia. Soilless culture: Prospects and challenges. Proceeding of the first Meeting of the

FAO Thematic Working Group of soilless Culture. 2 Septembre 1999. Halkidiki, Grèce.

26. Klougart A (1983). Substrates and nutrient flow. Acta Hort., 150: 297- 313.
27. Lemaire F, Dartigues A, Rivieère LM, Charpentier S (1989). Cultures en pots et en conteneurs. Principes agronomiques et applications. Edition INRA. Paris et PHM. Revue horticole, p. 181.
28. Leonardi C (2006). Substrates. 26-30. ECOPONICS-SOP project, pp. 27-30.
29. Leoni S, Ledda L (2004). The consequences of the methyl bromide ban on vegetable protected crops in sardinia. International work shop: "The production in the greenhouse after era of the methyl bromide". Comiso, Italy, 1-3 Apr. 2004.
30. Lieten F, Longuesserre J, Baruzzi G, Lopezmedina J, Navatel JC, Krueger E, Matala V, Paroussi G (2004). Recent situation of strawberry substrate culture in Europe. Acta Horticult. 649: 193-196.
31. Loveys BR, Schurwater F, Pons TL, Fitter AH, Atkin OK (2002). The relative importance of photosynthesis and respiration in determining growth rate. Plant cell Environ. 25(1): 975-998.
32. Maloupa E (2000). Alternative crops and growing systems for vegetables under protected cultivation in Mediterranean conditions. National Ag. Research Foundation of Greece. 4 Oct 2000.
33. Maloupa E (2002). Hydroponic systems. In Savvas, D. and Passam, H.(eds). Hydroponic Production of Vegetables and Ornamentals. Embryo Publications, Greece,
34. Marcelis LFM, Heuvelink E, Goudrian J (1998). Modelling biomass production and yield of horticultural crops: A review. Sci. Horticult., 74 (1): 83-111.
35. Metin-Sezen A, Celic S, Yazar G, Yalcim A, Mendi Y, Sahinler S, Tekin S, Gencel B (2006). Effects of drip irrigation management and different soilless culture on yield and quality of tomato grown in a plastic house. Pakistan J. Biol. Sci., 9(4): 766-771.
36. Morard P (1995). Les cultures végétales hors sol. I. N. P. T, France, p. 303.

37. Olympious CM (1992). Soilless media under protected cultivation rockwool, peat, perlite and other substrates. Acta Hort., 401: 443- 451. Radhouani et al. 585

38. Ould Djeh TK, Dali N, Bettaieb T, Ben Salah A (2006). Influence métabolique du CO2 atmosphérique sur la tomate cultivée en milieu saumâtre. John Libbey Eurotext. Cahiers d'études ET de recherches francophones / Agricultures, 15 (5): 441-447.

39. Özeker E, Eltezr Z, Tüzel Y, Gül A, **Önal** K, Tanrýsever A (1999). Investigations on the effects of different growing media on the yield and quality of strawberries grown in vertical bags. Acta Hort., 486: 409-414.

40. Papadopoulos AP, Hao X, Tu JC, Zheng J (1999). Tomato production in open or closed rockwool culture systems with NFT or rockwool nutrient feedings. Acta Hort., 481: 89-96.

41. Park SG, Lee BS, Chung SJ (1999). Effect of substrates on growth and fruit quality of "Mudeungsan" watermelon grown in hydroponics. J. Korean Soc. For Hort. Sci., 40: 419-424.

42. Qaryouti MM, Qawasmi W, Hamdan H, Edwan M (2006). Influence of NaCl salinity stress on yield, plant water uptake and drainage water of tomato grown in soilless culture. ISHS International Symposium on Protected Culture in a Mild-Winter Climate. Advances in soil and soilless cultivation under protected environment. February 19-24, 2006. AGADIR, MOROCCO.

43. Radhouani A, El Bekkay M, Ferchichi A (2008). The actual situation of the geothermic sector in the South of Tunisia. International Symposium of Biotechnology. Sfax – Tunisia May 4th – 8th, 2008.

44. Resh HM (1978). Hydroponic Food Production. A definitive guidebook of soilness food growth methods. Woodbridge Press Publishing Company, pp. 10-34.

45. Resh HM (1989). Hydroponics foodproduction. Woodbridge Press publishing company. Santa Barbara, California 93160, U.S.A., pp. 462.

46. Robin P (1998). Horticulture sans sol : Histoire Et Actualité. Cahiers d'économie et sociologie rurales, n° 46-47.

47. Savvas D (2003). Hydroponics: A modern technology supporting the application of integrated crop management in greenhouse. Food Agric. Environ., 1(1): 80-86.

48. Schröder FG, Lieth JH (2002). Irrigation Control in Hydrophonics. In.'Hydrophonic Production of Vegetables and Ornementals'. Ed. by D.Savvas and H.Passam. Embryo Publications Athens, Greece, pp. 263-298.
49. Schwarz M (1994). Soilless culture management. Springer-Verlag. 196 pp.
50. Sheikh BA (2006). Hydroponics: Key to sustain agriculture in water stressed and urban environment. Pak. J. Agri., Agril. Eng. Vet. Sc., 22(2): 53-57.
51. Tzortzakis NG, Economakis CD (2008). Impacts of the substrate medium on tomato yield and fruit quality in soilless cultivation. Hort. Sci. (Prague), 35(2): 83-89
52. Verdonck O (1975). Horticultural substrates. Wageningen,International Agricultural Centre, p. 88.
53. Verdonck O, De Vleeschauwer D, De Boodt M (1981). The influence of the substrate to plant growth. Acta Hort., 126: 251-258.
54. Xu HL, Gautier L, Cosselin A. (1997). Greenhouses tomato photosynthetic acclimation to water deficit and response to salt accumulation in the substrate. Japan Soc. Hort. Sci., 65(4): 777-784.
55. Yetisir H, Sari N, Aktas H, Karaman C, Abak K (2006). Effects of different substrates on plants growth, yield and quality of watermelon grown in soilless culture. American-Eurasian J. Agric. Environ. Sci., 1(2): 113-118.

Chapter 4

Nature and Causes of Land Degradation and Desertification in Libya: Need for Sustainable Land Management

Ali Mansour Ali Saad, Noresah Mohd Shariff, and Sanjay Gairola

Geography Programme, School of Distance Education, Universiti Sains Malaysia 11800 Minden, Penang, Malaysia

ABSTRACT

In this article, the characteristics of desertification, causative factors and efforts to combat desertification in Libya are presented. Also, the need to mainstream sustainable land management into land use planning is highlighted which could provide a firm basis for future policy decisions to combat desertification. Libya is characterized by a desert type environment and more than 95% of the country is desert or semi-desert. The processes of desertification have been aggravated by human activities coupled with climatic conditions. This intensification of desertification is believed to affect regional as well as global climate. Currently, desertification is one of the main environmental issues in Libya affecting environment and its resources. Among others, overexploitation of natural resources, inappropriate land use planning, insufficient water resources etc. are the main factors escalating the process of desertification and deteriorating environmental quality. Mainstreaming sustainable land management into land use planning has been considered a viable solution to moderate the effects of desertification and rampant usages of natural resources. However, this requires quantifying the severity of desertification by means of identified causative factor. Development of indicators of desertification process by means of using time-series satellite data coupled with ancillary data need to be attempted to establish a monitoring system to manage desertification in more sensitive areas. Moreover, environmental and livelihood implications of increasing desertification need to be addressed in order to promote regional economic sustainable development.

INTRODUCTION

Desertification is a serious environmental problem, involving the degradation of land in arid, semiarid, and dry sub-humid areas. It is caused primarily by human activities and climatic variations (UNEP, 1991; Zhang et al., 2008). According to MEA (2005), desertification is caused by a combination of factors that change over time and vary by location. Due to its particular geographical position and extreme climatic variations, Libya is experiencing a serious problem of land degradation and desertification. Climate change predictions for North Africa including Libya shows rising temperatures with potentially grim impact on the regions already stressed resources including water and food (IPCC, 2007). Importantly, the rate of dryland degradation and possible desertification may be further exacerbated by global climate change (MEA, 2005). Urbanization coupled with the loss of fertile soils, overexploitation of water resources, overgrazing, destruction of natural vegetation and rapid land use change are important reasons for the environmental problems in Libya. Despite the vast area of the country, most of this area lies within the warm desert climate that prevails in most of the northern part of the African continent where the desert covers 98% of its territory (Bin-Khayal, 1995). The only exception is a narrow coastal strip extends along the Mediterranean Sea and some mountainous areas in the north and south, where rainfall in sufficient quantities is enough for the growth of natural vegetation and cultivation of certain crops and fruit trees.

In spite of the small population of the country, the concentration in the northern region caused mankind pressure on the area, especially in marginal areas which resulted in serious environmental problems such as degradation and lowering of the groundwater table, seawater intrusion into fresh water, degradation of agricultural land and low productivity in many crops (El- Tantawi, 2005). All these factors largely contribute to the problem of desertification in Libya. It is pointed out that desertification; land degradation and drought deprive people of food and water and force millions to leave their homes. Agricultural areas have a potentially huge significance for food security as well as sustaining livelihood. Desertification of agriculture areas or pushing agriculture from non-desert to desert, not only agriculture itself may be doomed, but also alternative uses of the desert, such as controlled grazing, conservation of biodiversity, recreation, eco-tourism and tourism, are excluded (Portnov and Safriel, 2004). Impairment of agricultural ecosystems due to desertification often leaves behind an ecosystem that ceased to provide environmental services, and sometime is even impossible to rehabilitate. In Libya, agricultural practices are reported opportunistic mainly due to the country's harsh terrain, climatic conditions and limited irrigation which pose stern constraint to agricultural production. The Government of Libya has also instituted a number of agricultural research projects to promote agricultural development in the country. Researchers have also warned that the temperature increases and shifting precipitation patterns due to climate change may lead to further nitrogen losses in arid ecosystems and make arid soils even more infertile and unable to support most plant life (McCalley and Sparks, 2009). Assessment of the current global status of desertification has shown that accurate hard data, which would allow it to be stated with some precision as to which degree and with what rate of desertification is taking place in various parts of the world, are still lacking (Mamdouh, 1999). In this article, the characteristics of desertification causes and efforts to combat desertification in Libya are discussed. The need to mainstream sustainable land management into land use planning is also highlighted which could provide a firm basis for future policy decisions to combat desertification.

LIBYA: A BRIEF OVERVIEW

Libya occupies a large area of the northern part of the African continent (Figure 1) with an area of about Saad et al. 13681 1,600,000 km^2 (Libyan General Planning Council, 2003) Libya is located between 18 and 33° north of the latitude and 9 and 25° east of the longitude. Its climate is mainly determined by contrasting Mediterranean and Sahara climates (El-Tantawi, 2005). The Mediterranean Sea is regarded as the border from the north with a coast that has a length of 1900 km stretching from east to west (Abu Luqmah, 1995). The population of Libya in 2006 is about 5,657,692 million people according to the general census of population (Libyan General Agency of Information and Documentation, 2006). The vast area of the country and the difficult natural environment makes it impossible to develop the bulk of the area. On top of that desert conditions, water scarcity, and harsh climatic conditions have created barriers which impeded the expansion of human activity and development. As a result, the cultivated and populated areas are small and limited in comparison with the total area of the country. The desert climate prevails in the largest part of the country, with the exception of only a narrow strip extends along the Mediterranean which includes the coastal plains and the Northern highlands where the Mediterranean relative cold and rainy in winter with two transitional seasons which are spring and autumn. However, most of Libya experiences long periods of aridity and very few humid periods. Temperature ranges between high to very high in the summer and moderate to cooler in the winter and reaches the highest during the summer in August, while the lowest is recorded during the winter season in January. The climate of Libya is dry and the rainfall is almost limited to a few months of the year (that is, October to March). About 90% of the annual rainfall occurs during the cold half of the year, and nearly half of the amount falls during the months of January and February (Emgaiili, 1995). Rainfall is the main feature of precipitation in Libya; nonetheless, it is very erratic and limited (El-Tantawi, 2005). Given this sporadic rainfall pattern, agriculture production in Libya is highly reliant on irrigation. Rainfall in Libya is very infrequent, thus limited dependence on rainfall for agricultural production. For example, only 32% of wheat, 38% of barleys, 25% of dates, 29% of pulses, 34% of vegetables and 24% of olives are produced under rain-fed conditions (FAO AQUASTAT, 2005). Hence, climatic conditions severely limit agricultural production in Libya. Average annual rainfall distribution in Libyan territory is presented in Table 1.

Libya like many Mediterranean regions, also experiences more drought and water deficits that play a major role in the process of desertification. Moreover, Abufayeda and El-Ghuelb (2001) reported that this deficit will undoubtedly increase in the future in response to the continuous population growth and corresponding increase in water requirements for domestic, industrial and agricultural purposes. Currently, the water demand exceeds the conventional water resources capacities markedly creating an urgent need for integrated water resources management with special focus on nonconventional water resources namely; sea water desalination and wastewater reuse (Wheida and Verhoeven, 2007). This situation of water deficit is becoming more problematic with rapidly increasing population and low rainfall. The high temperatures and low precipitation resulted in low vegetation cover, increased soil erosion and land degradation at large scale and put the region on highest desertification risk. The annual rate of evaporation is about 1700 mm near the sea and 6000 mm in the central and southern regions (Libyan General Planning Council, 2003). Groundwater is the main source of water supply in Libya, and represents more than 97% of the total water consumed for different purposes. Groundwater in Libya is classified into renewable groundwater and non-renewable underground water. Safe water extraction is estimated about 3000 million cubic meters per year (Libyan General Planning Council, 2003) and located in five major water basins. Table 2 shows water situation in Libya from 1990 to 2025. In general, Libya is regarded as one of the areas with low density of natural vegetation due to lower annual rates and irregular distribution of rainfall. Additionally, overgrazing has also increased the process of destroying natural vegetation cover and land degradation in many parts of the country. Eventually, this has increased the severity of desertification problem and environmental degradation.

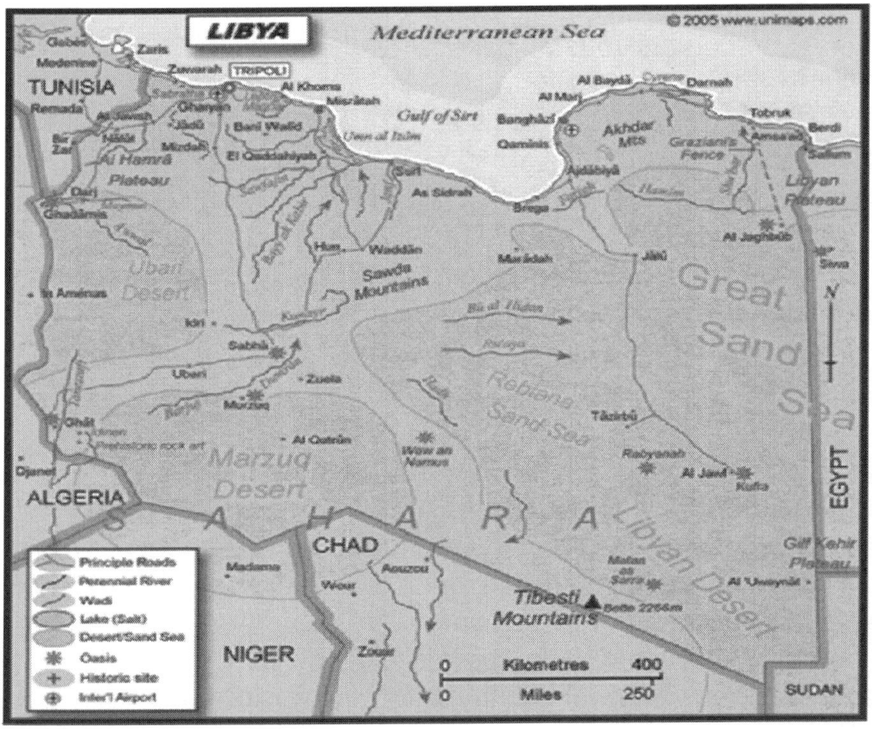

Figure 1: Distribution of desert in Libya.

Table 1: Area as per Aridity zones of Libya

Aridity zone	Average annual rainfall (mm)	Land area (000 km2)	% Total area
Very dry	<50	1589	90.8
Dry	50 — 200	130	7.4
Semi-arid	200 — 400	26	1.5
Sub-humid	>400	5	0.3
Total		1750	10

Table 2: The water situation in Libya (1990 to 2025)

Sector-wise demand	1990	2000	2010	2020	2025
Required (million m3)					
Agriculture	4275	4800	5325	5850	6640
Drinking	408	647	1015	1512	1759
Industry	74	132	236	422	566
Total	4757	5579	6576	7784	8965
Available					
Renewable water	500	500	500	500	500
Non-traditional sources	105	127	155	188	208
Artificial river		1642	2226	2226	2226
Total	605	2269	2881	2914	2934
Deficit	4152	3310	3695	4870	6031

CAUSES AND CONSEQUENCES OF DESERTIFICATION

In recent decades, Libya has witnessed enormous development in various fields, particularly in the sectors of agriculture and industry. However, such development has put negative effects on local ecosystems, especially in sensitive and fragile areas due to change in production patterns, and the need to provide food requirements for the growing population. Consequently, this lead to the intensification of pressure on already limited natural resources, and thus escalating land degradation and desertification problems. There are several natural and human factors that cause land degradation in Libya (Ben- Mahmoud et al., 2003). The natural factors include changes in climate, mainly rain, wind and temperature. Climate varies from one setting to another in response to changes in external and internal influences in the air system (Emgaili, 1993). There have been significant climate changes through geologic time on the African continent, in which wet and dry ages succeeded. Dry ages led to the emergence of the Sahara, and the current climate of the area is a continuation of the dry climate, which began since the emergence of the Sahara, with a general tendency towards more droughts in the past years. Therefore, the frequent wind and water erosion are symptom of land degradation and desertification in Libya (Emgaili, 2003). Although, climatic factors play an important role in the spread of the phenomenon of desertification, human activity causes the deepening of the effects of drought on environmental resources and human beings. Many human factors are combined contributing to the deterioration of environmental conditions and the occurrence of desertification in Libya. The increasing pressure of population on natural resources (example water, soil, vegetation), poor management and over exploitation of natural resources by entities or individuals, mainly led to desertification. Anthropogenic factors causing deterioration of soil and vegetation in Libya include: (i) overexploitation of water resources (ii) rangeland conversion cropland (iii) deforestation and the removal of natural vegetation (iv) over-grazing in marginal areas (v) misuse of the soil on arable land, and urbanization (El-Tantawi, 2005; Emgaili, 2003; Libya General Planning Council, 2003).

Desertification results in a variety of negative environmental, economic and social impacts, directly or indirectly (Nahal, 1987). Consequently, the productivity of natural pastures, forest and agricultural land is decreasing. In periods of successive drought, the problem is considered acute and a significant decrease in the production per hectare of grain in the dry and marginal areas in the Arab world is observed. Moreover, deterioration of the pastoral environment causes a decline in the productivity of livestock and thus, reduces the productivity of meat and milk. Desertification is accompanied by deterioration in soil fertility, change in its physical and chemical properties and vulnerability to water erosion (Arab League, 2003).

This may cause a decrease in the volume of agricultural resources loss and decrease of the area of arable land, which ultimately has negative socio-economic consequences. The shrinking of forest lands and natural pastures due to deforestation leads to a vicious circle in which there is an increase in the number of animals beyond the carrying capacity, leading to the importing of animal feeding from abroad, and severe economic losses. That impacts on national income (Nahal, 1987). Desertification also leads to many environmental impacts. A major consequence of desertification at the local and global level is the reduction in biodiversity, since it contributes to the destruction of the habitats of animal and plant species and micro-organisms (Abahussain et al., 2002). Biodiversity loss is also expected to encourage the genetic erosion of local livestock and plant varieties and species living in fragile ecosystems. Libya is no exception.

EFFORTS TO COMBAT DESERTIFICATION

Since the early sixties, serious measures have been taken to combat desertification in Libya at the best possible way (Bin Mahmoud, 2000). These measures were part of a broad policy in the framework of National Plan for Agricultural Development which takes into account the objectives of local development on the one hand, and harsh environmental conditions prevailing in the country on the other. These measures include curbing sand dunes, establishment of windbreaks, reforestation of fallow forest land, establishment of terraces to combat soil erosion, preservation of rain water on sloping agricultural land, and follow the special agricultural cycle to maintain soil fertility, especially in the areas of cultivation of grain, as well as the protection and improvement of natural pastures (Nahal, 1987). Literature shows that in spite of the success of some attempts to achieve the desired objectives, others did not meet the same success due to the lack of relevant laws and legislations concerning the protection of the environment (Libyan Department of Urban Planning, 2005). However, adopting better land use management practices could slow down the desertification process. Libya has adopted a lot of measures and actions to reduce the desertification during the past four decades. Major desertification controlling strategies can be summarized in the implementation of a range of diverse projects by government in many areas (that is, forest, pastures, sand dune fixation, soil and water conservation, resistance to erosion and integrated agricultural development). Libya has given a great attention to development and conservation of water resources by adopting a short term and long term strict policy to prevent a serious drain of water resources. There are many other actions that contribute to the protection of soil and maintain soil fertility, such as, following the method of crop rotations to maintain soil fertility, especially in the areas of expansion in the cultivation of grain; reforestation of waste lands; protecting and improvement of natural pastures; the creation of economic non-agricultural activities, especially in areas threatened by agricultural depletion or degradation of quality (Libyan Department of Urban Planning, 2005). Libya has paid a great attention for pastures improvement and development as well as the establishment of pastoral projects based on scientific methods to ensure the preservation of ecological balance. Wise attempts have been made to regulate grazing and rehabilitation of degraded lands by a number of programs. Considerable efforts are made to reforest lands threatened by erosion and desert encroachment in order to maintain the natural ecological balance. These include protecting land and soil, and providing some of the requirements of forest products, combating unjust cutting of forests and developing the existing forests to increase their productivity, establishment of nurseries to meet the needs of the citizens for seedlings, supporting small forest planting, windbreaks, and protective barriers, as well as meeting the needs of public projects when establishing new forests, and filling the requirements of afforestation and roads are amongst the important efforts. Moreover, determining access rights, and preventing the cutting of some species, protecting them from insect damage, and the organization of grazing in natural forests are among significant initiatives (Libyan General Planning Council, 2003). For reforestation of degraded and deforested areas to compensate the losses, the country has witnessed campaigns of tree planting from the 1990s onwards, including most of the desertified and degraded lands for the purpose of balancing natural ecosystem, land rehabilitation establishment of national parks and wildlife reserves to maintain biodiversity and prevent extinction of rare animals and plants, and issuing legislations and laws for the protection, development and investment of forests. Failures of resource management policies are aggravated by overgrazing, overexploitation of water and land resources, and over-cultivation of marginal lands, deforestation, and the use of inappropriate technologies.

ESTABLISHMENT OF LEGAL PRINCIPLES

Establishment of a global network of national, regional and international institutional and technical facilities for current operational assessment and continuous monitoring of desertification has been effectively debated in many environmental research forums. A pressing and constant need was felt to develop laws and legislative enactments to serve as an appropriate framework for the efforts made to combat desertification and achieve the objectives and requirements of sustainable development. In 1977, the United Nations Conference on Desertification (UNCOD) adopted a Plan of Action to Combat Desertification (PACD). Also, the United Nations Conference on Environment and Development (UNCED) in Rio de Janeiro in 1992, called upon the United Nations General Assembly to establish an Intergovernmental Negotiating Committee (INCD) and Convention to Combat Desertification in affected countries, particularly in Africa. Libya has ratified important international agreements on environment, like biodiversity, climate change and desertification. There are several pieces of legislation to protect the environment, natural resources including agricultural, pastures, urban development, rationing water and soil use, protection of renewable and nonrenewable natural resources etc. Few of these legislations are included in Table 3. In light of these and other laws and legislations, it was possible to achieve many of the objectives of combating desertification and halt desert encroachment, and maintaining the ecological balance while conserving biodiversity, both in terms of legislative or operational terms. However, the enactment of policy, laws and legislations need to be assessed properly through appropriate framework.

Table 3: Important Law and Legislations regarding environment and natural resources in Libya

Law and Legislation	Main objective
Law 15 of 1992	Protection of agricultural lands, pastures and forests and converting them to irrigated agricultural lands.
Law 72 of 1988	Establishment of the Arab Center for Desert Research and development of desert communities.
Law 15 of 1984	Protection of animals and trees, and to prevent hunting wild animals, and the prevention of trees cutting because of urban expansion.
Law 1 of 1983	Agricultural inspection

Law 790 of 1982	Organization of drilling operations and the preservation of water sources.
Law 7 of 1982	Protection of the environment
Law 5 of 1982	Protection of pastures and forests
Law 827 of 1980	The establishment of the General Authority for Scientific Research and its bodies specialized in various fields.
Law 46 of 1972	Protection of shrub land.
Law 26 of 1972	The establishment of a public board of water responsible for proposing public policies and legislations concerning water and follow up their implementation, as well as overseeing the follow-up projects relate to water abstraction, digging wells and methods of using them.

CONSTRAINTS AND PROSPECTS

Despite the efforts and achievements made so far, the factors and causes, manifestations and harmful effects of desertification need special attention in Libya. Researchers, natural resources managers and conservationists highlighted a number of obstacles and challenges that need to be confronted by all possible means including: (i) Scarcity of water resources and successive seasons of drought, which limit the success of agricultural projects, land reclamation and cultivation (ii) Increasing the level of food deficit as a result of rapid population growth, and the widening gap between production and consumption rates, and the continued expansion of urban areas which causes damage to agricultural production (iii) Erroneous practices of population and unorganized exploitation of natural resources without paying any attention to preserve them (iv) Incorrect application of the provisions and legal regulations and legislation governing relations between citizens and natural and environmental resources (v) The absence of a comprehensive database of natural resources in terms of importance and restriction for their uses, and (vi) Weakness of skilled manpower and specialized staff in the field of protection of natural specialized staff in the field of protection of natural resources and combating desertification (Libyan Department of Urban Planning, 2005). Climate change will result in extreme weathers for the region affecting both the intensity and frequency of climate factors such as temperature, precipitation, rainfall and droughts (IPCC, 2007). In Libya, water scarcity is endemic and changes in precipitation could pose a strain to freshwater resources, vegetation and desertification (El-Tantawi, 2005). Currently, many areas became prone to land desertification and the situation even may go worst if climate change and human activities continue to trigger land degradation in the country. Climatic factors contribute to both desertification and the scarcity of water; and no doubt global warming is exacerbating these problems. As far as possible solutions to desertification is considered, many researchers highlighted that these problems can be tackled by re-planting tree in shelter belts as a major afforestation programmes, planting grasses to stabilize the soil and halt wind and water erosion, and obviously adopting proper crop rotation and use of manure as a fertilizer, as a good farming practices. Rehabilitating all of the desertified land in the world is not economically profitable. However, if people stop misusing the land and overexploitation of resources beyond the carrying capacity, desertification can be controlled at large extent and management practices can be implemented more effectively.

CONCLUSIONS

Sustainable development and sustainable living includes social, economic and environmental considerations and dimensions. Desertification and climate change modify all these dimensions, and therefore alter potential development goals and efforts (Seely et al., 2008). As the degree and types of desertification varies from one country to another within the region, land management has been acknowledged to be extremely important in desertification prone areas. Understanding the desertification process more explicitly requires detailed knowledge about where desertifications occur, spatial as well as geographic gradients. Climate change is expected to decrease water availability and water quality, increase droughts, floods and salinity leading to decrease in soil fertility and loss of vegetation threatening food security. The increasing pressure on resources by people in addition to the severe climate and low soil fertility has rendered the ecosystems even more fragile, and in some places their renewal is jeopardized. It is reported that the progression of desertification is making the recovery of vegetation on cleared and abandoned land impossible, so human assistance through soil preparation, use of fertilizer and watering during regeneration and careful monitoring is needed. The remote sensing (RS) technology with its multi-temporal, multi-spectral, synoptic and repetitive coverage can provide valuable information on intensification of desertification, real magnitude of desertification and its changes over time. To date, there are few studies of desertification monitoring in Libya using RS/GIS tools (Abahussain et al., 2002; Emgaili, 2003, El-Tantawi, 2005), but comprehensive studies are still awaited. To provide useful information on the prevailing desertification situations in Libya, areas having exceptionally high magnitude of desertification need to be delineated with the help of RS/GIS tools. These geospatial tools could help in tracking landform changes to create and update maps for managing land utilization effectively and sustainably.

REFERENCES

1. Abahussain AA, Abdu A S, Al-Zubari WK, El-Deen N A, Abdul-Raheem M (2002). Desertification in the Arab Region: analysis of current status and trends. J. Arid Environ., 51: 521–545.
2. Abufayeda AA, El-Ghuelb MKA (2001). Desalination process applications in Libya. Desalination, 138(1-3): 47-53.
3. Abu Luqmah M A (1995). Aljamahirya: a study in geography. First edition. Aljamahirya Home for Publication and Distribution, Sirt, Libya, p. 17.
4. ACSAD (The Arab Center for the Studies of Arid Zones and Dry Lands) (2004). State of Desertification in the Arab World, Damascus.
5. Al-Balooshi ASS (2003). Desertification in Al-Batinah Plain in Sultanate of Oman. Unpublished doctoral dissertation, University of Jordan, Amman, Jordan.
6. Arab League (2003). A study on indicators for monitoring desertification in the Arab World, Khartoum, Sudan. Arab Organization for Agricultural Development.

7. Ben-Mahmoud R, Mansur S, AL-Gomati A (2000). Land degradation and desertification in Libya. Land Degradation and Desertification Research Unit, Libyan Center for Remote Sensing and Space Science, Tripoli, Libya.
8. Ben-Mahmoud R, Mansur S, AL-Gomati A (2003). Land degradation and desertification in Libya. In: Alsharhan AS, Wood WW, Goudie AS, Fowler A, Abdellatif EM (eds) Third Millennium. Swets and Zeitlinger Publishers. Lisse, the Netherland, pp 339-350.
9. Bin Khayal A S (1995). Agriculture and Livestock chapter, Al-Jamahirya: a study in geography. 1st edition, Aljamahirya home for publication and distribution, Sirt.
10. El-Tantawi A M M (2005). Climate Change in Libya and Desertification of Jifara Plain: Using Geographical Information System and Remote Sensing Techniques. Dissertation, the Johannes Gutenberg University, Mainz, German.
11. Emgaili E (1993). Climate changes and its impact on human and the environment. 1st issue, the University Magazine.
12. Emgaili E (1995). Climate changes, Al-Jamahirya: a study in geography. 1st edition, Aljamahirya home for publication and distribution, Sirt, Libya.
13. Emgaili E (2003). Risks of drought and desertification and phenomena associated with both of them. 1st edition, Candles culture for print, publishing and distribution, Zawia, Libya.
14. Fadel M A, Abu Luqma A (1995). The water resources chater, Al-Jamahirya: a study in geography. In: Abulugma A, Saad A (eds) 1st edition, Aljamahirya home for publication and distribution, Sirt, Libya.
15. FAO AQUASTAT database (2005). FAO's Information System on Water and Agriculture. Food and Agricultural Organization of the United Nations, Rome, Italy.
16. Libyan Department of Urban Planning; General Authority for Infrastructure and Urban development (2005). Draft plans thirdgeneration, the scope of Tripoli schematic, Report of the environmental factors.
17. IPCC (2007). Climate Change 2007. IPCC Fourth Assessment Report Working Group I Report "The Physical Science Basis". Cambridge University Press, New York.

18. Libya General Planning Council (2003). Report about agricultural policies. Tripoli,
19. Libya. Libyan General Agency of Information and Documentation (2006). The final results of general census of population. Tripoli.
20. MEA (2005). Ecosystems and Human Well-being: Desertification Synthesis. World Resources Institute, Washington, D.C. pp.26.
21. Mamdouh N (1999). Assessing Desertification and Water Harvesting in the Middle East and North Africa: Policy Implications. ZEF – Discussion Papers on Development Policy No. 10, Center for Development Research, Bonn, July 1999, pp. 59.
22. McCalley CK, Sparks JP (2009). Abiotic gas formation drives nitrogen loss from a desert ecosystem. Science, 326(5954): 837-840.
23. Nahal I (1987). Desertification in the Arab world, Arab Development Institute, Beirut, Lebanon.
24. Portnov BA, Safriel UN (2004). Combating desertification in the Negev: dryland agriculture vs. dryland urbanization. J. Arid Environ. 56: 659– 680.
25. Seely M, Dirkx E, Hager C, Klintenberg P, Roberts C, von Oertzen D (2008). Advances in desertification and climate change research: Are they accessible for application to enhance adaptive capacity?. Global Planet Change, 64: 236–243. Saad et al. 13687
26. UNCED (1992). Managing fragile ecosystems: combating Desertification and Drought (Agenda 21, Chapter 12). United Nations Conference on Environment and Development, (UNCED), Rio de Janeiro, Brazil.
27. UNEP (1991). Status of desertification and implementation of the United Nations plan of action to combat desertification. Nairobi, Kenya.
28. Wheida E, Verhoeven R (2007). An alternative solution of the water shortage problem in Libya. Water Resour. Manage. 21:961-982.
29. Zhang Y, Chen Z, Zhu B, Luo X, Guan Y, Guo S, Nie Y (2008). Land desertification monitoring and assessment in Yulin of Northwest China using remote sensing and geographic information systems (GIS). Environ. Monit. Assess. 147: 327–337.

Chapter 5

Distribution and Regeneration Status of Vitex Payos (Lour.) Merr. in Kenyan Drylands

James Munga Kimondo[1], Jacob Godfrey Agea[2], Clement Akais Okia[3], Andrew Woiso Dino[4], Refaat Atalla Ahmed Abohassan[5], Jackson Mulatya[1], and Zewge Teklehaimanot[6]

[1]Kenya Forestry Research Institute, P. O. Box 20412-00200, Nairobi, Kenya

[2]College of Agricultural and Environmental Sciences, Makerere University, P. O. Box 7062 Kampala, Uganda

[3]ICRAF Uganda Country Office P. O. Box 26416 Kampala, Uganda

[4]Department of Biological Science, Sokoine University of Agriculture, P. O. Box 3038, Morogoro, Tanzania

[5]Faculty of Meteorology, Environment and Arid Land Agriculture, King Abdulaziz University, P. O. Box 80208 Jeddah, 21589 Saudi Arabia

[6]School of Environment, Natural Resources and Geography, Bangor University, Bangor-Gwynedd LL57, UK

ABSTRACT

We investigated the population structure and regeneration status of Vitex payos (Lour.) Merr in Kenyan drylands. The study quantified the spatial distribution pattern of V. payos tree populations in their natural range; and assessed their regeneration status to determine the stability of the populations Woodlands and farm inventories were conducted in Mbeere, Mwingi and Kitui districts of the Eastern Province of Kenya. The nearest-neighbour sampling method was used to determine tree density and distribution of V. payos in the study sites. The number of seedlings and saplings were counted. The diameters, crown diameter, and tree heights of sampled trees were measured. These morphological parameters were summarized on per hectare basis. The patterns of distribution of V. payos trees showed an aggregation of trees on farms and bushes. Tree densities ranged from 1.6 on farmlands to 20.3 trees per ha in the woodlands. The expected mean distances between nearest neighbouring trees were higher than the observed values on all sites, confirming that V. payos trees were more aggregated than randomly dispersed. The mean tree heights varied from < 5 m to > 9 m cross study sites. The sampled populations were dominated by trees (55%) within the range of 10 to 20 cm dbh. Highhest numbers of seedlings from all origins were recorded in the bushes in Kitui and Mbeere districts 101 and 78 seedlings per hectare, respectively. Sapling population densities were generally low.

INTRODUCTION

Past agricultural activities are some of the major factors that have influenced the present day species composition cultivation of food crops, planting of exotic tree species and overstocking of livestock that interfere with the regeneration of natural vegetation through continuous trampling and/or browsing of young plants as well as the reproductive parts of adult plants - flowers and fruits (Turner et al., 1998). In particular, the planting of exotic species for different end-use products including fruits currently poses a major threat to indigenous species such as *Vitex payos*, yet many of these indigenous species are considered slow growers and therefore, often not preferred by farmers for on-farm planting (Dunn, 1991). The increasing human population exploiting a given environment, land fragmentation and climate change also contribute to the overall decline in indigenous vegetation (Wanyonyi, 2012; Oroda, 2011; Morris, 2010).

V. payos like other indigenous fruit trees are commonly found scattered on farms in Kenya (Jama et al., 2008; Ekesa et al., 2009). Small scale farmers retain naturally regenerated *V. payos* trees on arable fields, grazing lands, homesteads and along farm boundaries for various purposes such as fruits, apiculture, timber, shade, soil amelioration, firewood among others (Jama et al., 2008). The species regenerates naturally by seed, root suckers and coppicing (Maundu et al., 1999; Mbora et al., 2008). Root suckers often develop from injured roots especially during land preparation for agricultural purposes (Mbora et al., 2008) or construction of soil conservation structures such as terraces. Due to heavy browsing of the regenerates, especially during the dry season, majority of them remain stunted for long periods. Regenerates established from seeds often suffer heavy mortality because of combined effects of browsing and drought before their root systems are fully developed (Luoga et al., 2004). On the contrary, regenerates that grow from root suckers and stumps benefit from already established root systems and therefore tolerate the extreme environmental vagaries with less threat of mortality (Kauter et al., 2003). Trees in woodlands are often small and suffer from the dry weather due to excess competition. This is unlike on the farms where trees are properly protected through weeding thereby reducing competition for both nutrients and water resulting in bigger trees. However, due to the principal purpose of the land is agricultural, only a few trees are retained. These trees also maintain their leaves for a longer period even after the on-set of the dry season.

The objectives of this study were therefore to: quantify the spatial distribution pattern of *V. payos* tree populations in their natural range; analyze the size distribution pattern of *V. payos* trees under different land uses; and assess the regeneration status of the species in order to determine the stability of the populations. It was hypothesized that the tree spatial distribution in the natural range of the species is not different among locations and land uses. Secondly, that the density of seedling, sapling and mature trees are each randomly distributed on farms and bushlands with their populations among the various natural vegetation types in Kenya(Turner et al., 1998). These activities include decreasing with increasing age, a J-inverse distribution, indicating stable regenerating populations.

MATERIALS AND METHODS

Study Area

The study was conducted in Mbeere, Mwingi and Kitui Districts which are located in the arid and semi-arid (ASAL) parts of Kenya. Mbeere is situated at between latitude 00°20′ and 00° 50′ S and longitude 37°16′ and 37° 56′ E with a mean annual rainfall of between 640 and 1100 mm, while Mwingi is situated at between 00°03′ and 01° 12′ S and 37°47′ to 38°57′ E and receives 600 to 1100 mm mean annual rainfall. Kitui District lies between 00°04′ and 03°00′ S and 37°45′ to 39°00′ E and receives a mean annual rainfall of between 500 and 1000 mm. The rains have bimodal distribution from March to May (long rains) and October to December (short rains) with peaks in April and November, respectively. Minimum and maximum temperatures in the study districts are fairly constant with values of 21.6 and 34.8°C in Mbeere, 22.8 and 37.2°C in Mwingi, 21.9 and 36.6°C in Kitui, respectively. These areas are predominantly under subsistence crop production practiced by low resourced farmers, who also rear large herds of livestock under free range grazing system. These areas are also characterized by low soil fertility and experience frequent droughts and famine.

Data Collection

Nearest-neighbour sampling method, in which the distance and direction of the 100 sampled trees were measured from a randomly selected individual tree to its nearest neighbour, was used to determine tree density and distribution of *V. payos* in the three study sites (Sparks et al., 2002). However, several trees were found to be each other's nearest neighbour and use of these distances twice could decrease sample variance (Donnelly, 1978). Furthermore, the trees at the edge of the imaginary plot could not have trees outside the fixed number as their nearest neighbours. To overcome the bias as a result of these two conditions, Donnelly's correction factor was used (Donnelly, 1978). Tree densities in circular plots could be estimated as $\rho = n/\pi r^2$, where ρ is tree density (trees m^{-2}), n is the sample size and r_d is the distance (m) of the uttermost tree from plot centre, assuming the area occupied by the n trees was circular. In each of the three study areas, two major types of land use, woodlands and on farm, were randomly selected for the survey of tree density, size class distribution and regeneration status of *V. payos*.

In each land use type, one mature *V. payos* tree (≥ 10 cm dbh) was randomly selected and used as the starting tree in all the three sites. Moving in a circular manner around the first tree and while increasing the radius, other 99 neighbour trees were identified, located and marked, keeping the area compact. All the tree positions were recorded using a GPS (GPS 12XL Garmin Olathe KS, USA). Bearings and horizontal distance to the nearest neighbour were also recorded between all trees in a plot using compass and meter tape. These recorded distances and bearings from the starting tree and from one tree to the others were used to estimate the spatial distribution and tree population density in each study site. The diameters at breast height (DBH), tree heights and crown diameters of all the 100 sampled trees were measured. DBH was measured with a diameter tape, height with a telescopic pole and crown diameter was derived as the average of two perpendicular crown diameters through the crown vertical projection. The area occupied by the 100 sampled trees was mapped graphically on a grid scaled at 50 × 50 m for the study of species regeneration. All the 0.25 hectare grids were allocated a number to allow random selection of plots to sample for seedlings and saplings. Ten percent of these 0.25 ha plots were randomly selected. Within these plots, the seedlings (≤ 1 m height) and saplings (> 1 m height but dbh ≤ 10 cm) were counted. Seedlings and saplings were classified as originating from seed, root suckers or stump coppices. They were distinguished as follows: those from seed generally had one stem at the ground level, those from root suckers were characterized by multiple stems at the base while those from stump coppice had visible stumps. When there were doubts regarding the origin of an individual seedling, root system was partially excavated to reveal the stump or the old root. The number of regenerates of each origin was used to calculate the density (stems ha^{-1}).

Data Analysis

The area occupied by the sample trees was more of a polygon with many sides as prescribed by outer trees. Consequently, the area of the plot, A in m^2, was obtained by tracing the coordinates of the outer edge trees and calculating the area of the triangles and rectangles. The lengths of the sides of the polygon prescribed by the outer trees were summed to give the perimeter *L* of the plot. The tree density was therefore computed as $\rho = n/A$. The mean distance between nearest neighbours was computed as $r_A = (\Sigma r_i)/n$, where r_i is distance of the ith tree from its nearest neighbour. The mean distances that would be expected, r_E, for each plot if the trees were randomly distributed and with the same tree density as the computed value and the variance, σr^2 were calculated using the Clark and Evans (1954) formula, $r_E = 1/2 \sqrt{(\rho)}$ and $\sigma r^2 = 0.0683/\rho$, where r_E is the expected mean distance to the nearest neighbour in an infinitely large population with a random distribution of density ρ - (where the unit of measurement used in the calculation of ρ must be the same as that used in measuring *r* – in this case m and m^2). However, with Donnelly (1978) correction factor, the corrected mean nearest neighbour distance (r_c) and the variance (σc^2) are,

$$rc = r_E + (0.0514 + 0.041/\sqrt{n})*(L/n) \text{ and } \sigma c^2 = \sigma r^2 + 0.037\sqrt{(1/\rho)}*(L/n),$$

Where L is the sample plot boundary length (as a polygon) and other variables as defined earlier. Assuming spatial randomness of the trees, the standard error of the corrected expected mean nearest neighbour distance (s_r),

$$S_r = \sqrt{(\sigma r^2 / n)}$$

such that the *p* value of deviation from the expected mean distance is obtained as (Campbell, 1996):

$$c=(r_A - r_C)/S_r$$

The ratio *R* of r_A/r_c was used as a measure of the extent to which the sampled populations deviated from a randomly distributed population of the same density. Where the trees are heavily aggregated, the *R* ratio tends to 0 as the distances between most neighbouring trees are close to or equal to 0 (Clark and Evans, 1954). At the other extreme, where trees are regularly and uniformly distributed, the *R* ratio is greater than 1 (Clark and Evans, 1954). Besides comparing the spatial distribution of trees between plots, descriptive statistics, non-parametric analyses and t-tests were calculated using SPSS for windows Version 16 to make comparisons between study sites (districts) and land uses (bush and farm). To compare individual districts where significant differences were found, Tukey's post hoc test was applied (Kleinbaum and Kupper, 1978; Dancey and Reidy, 2007). The tree morphological parameters of height, diameter at breast height and crown diameter were summarized on per hectare basis

and checked for normality using Levene test (Dancey and Leidy, 2007). Trees were assigned into dbh classes (10 ≤ 19.9 cm, 20.0 ≤ 29.9 cm, 30.0 ≤ 39.9 cm, 40.0 cm and above); height classes (≤ 5.00 m, 5.01 ≤ 7.00 m, 7.01 ≤ 9.00 m, and 9.01 m and above), and crown diameter classes (≤ 3 m, 3 - 5 m, 5 - 7 m, 7 - 9 m, and 9 m and above). Populations were compared using analysis of variance (ANOVA) at 95% confidence interval. Tukey's post hoc test allowed pair wise comparison of the mean heights, dbh and crown diameters of individual populations. To determine those populations that were actually different, a Mann Whitney U test was used to compare the two land uses while a Kruskal Wallis test was used to compare the districts. Seedlings and saplings were summarized on a per hectare basis from the 0.25 ha plots and ANOVA done to compare the different seedling origins while the land uses were compared using Student t test. Histograms were presented to show the regeneration trend among seedlings, saplings and trees for each population.

RESULTS

V. payos Tree Population Structure

Spatial distribution patterns of mature *V. payos* trees showed that they were not uniformly distributed across their natural range (Figure 1). Trees were aggregated both on cultivated farms as well as in bushlands. Tree densities in the study areas ranged from 1.6 trees per hectare on cultivated land in Mwingi District to 20.3 trees in bush/fallow land in Kitui District (Table 1). Tree densities were higher in bushlands than on cultivated land in all districts. The expected mean distances between nearest neighbouring trees were higher than the observed values on all sites under the two land uses, an indication that trees in these areas were more aggregated than randomly dispersed. This is further supported by the *R* ratio values of the three villages which are less than 1 and the negative standard variate of the normal curve (c) (Table 1). Distribution of trees was positively skewed as there were few isolated trees sampled that were far from the others (Figure 1 and Table 1).

Comparison of the six populations of *V. payos*, gave an *F* value of 5.97 indicating that the six populations were significantly different at $p = 0.05$. Seperation of the means revealed significant Chi-square values of 152.605 for districts and 156.799 for study plots both with an associated probability value of $p < 0.001$. Trees in Mbeere were relatively more dispersed while trees in Kitui were more aggregated (28.28 ± 1.60 m for Mbeere; 17.26 ± 1.74 m for Mwingi; and 6.29 ± 0.57 m for Kitui trees). In the study plots, the mean nearest neighbour distance ranged from 5.10 ± 0.7 m in on farm land in Kitui to 33.56 ± 2.47 m in farm lands in Mbeere District. There were no differences between combined data for each land use from the three districts with *Z* value of -0.505 and *p* value of 0.613.

At 95% confidence interval, the population structure of the *V. payos* in Kitui District, irrespective of land use, were not significantly different.

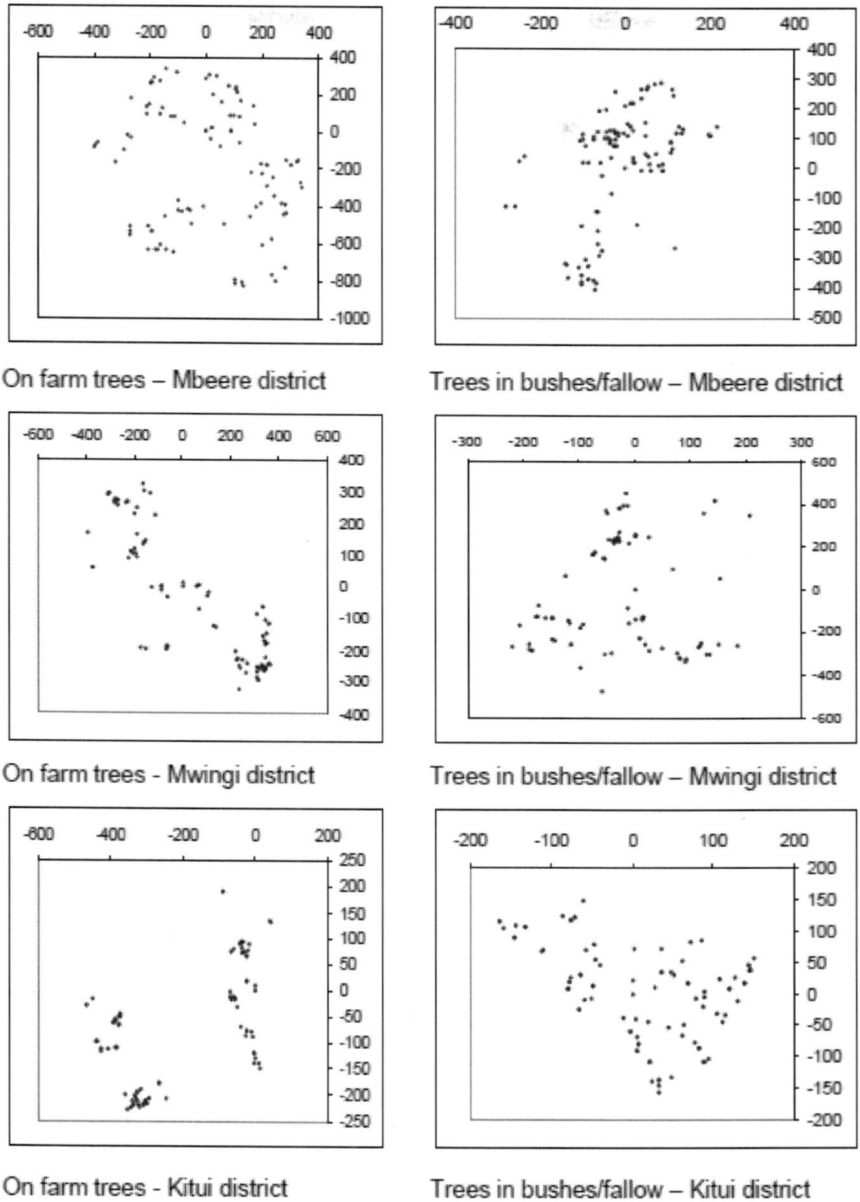

Figure 1: Spatial distribution pattern of mature *V. payos* trees on arable land and bushland in villages in Mbeere, Mwingi and Kitui Districts, Kenya (distances are in metres, 0,0 represents the center tree).

Table 1: Comparison of *V. payos* tree population densities and spatial distribution in arable and bushland in Mbeere, Mwingi and Kitui districts of Eastern pr ovince Kenya

		Mwingi		Kitui		Mbeere
Statistics	On farm	Bushesfiaflow	On farm	BushesWallow	On tam	Bushes/ fallow
Area =pied by sample trees as a polygon (m2)	283119.5	267361.5	137157	49275.5	610279	210422.5
Number of trees sampled (n)	100	100	100	100	100	100
Sample population density (ρ)	0.000353	0.000374	0.00073	0.00203	0.000164	0.000475
Sample bee density per hectare (m)	3.53	174	7.29	20.29	184	4.75
Total distance between nearest neighbour (Σ ri)	1386.56	2045.43	509.86	728.11	3356.32	2279.46
Mean distance between nearest neighbour (rA)= (Σ ri)/n	1186	20A5	5.10	7.28	33.56	22.79
Skewness of dishibuticri of mean nearest neighbour	1.74	1.94	1.82	1.01	0.72	2.23

Expected mean distance between nearest neighbour in a large randomly distributed population of ρ density (rE =1/2 $\sqrt{\rho}$)						
Length of the sample plot boundary L	2661	25.85	18.52	11.10	39.04	22.94
Corrected expected mean nearest neighbour distance rc	2132	2204	1470	925	2892	1799
rc= rE+(0.05140.041/4nr(Un)	27.79	25.99	19.34	11.61	40.65	23.94
R ratio (measures the degee to which observed distntuticn deviates from a randomly elsbibuted population of same density), (R = (ra/rc)) art= 0.2613614 (n. p): standard error of mean *stance to	0.50	0.79	0.26	0.63	0.86	

$\rho_{rE}= 0.26136/(\sqrt{n.p})$: standard error of mean distance to nearest neighbour in a randomly distributed population of density ρ.	1.39	1.35	0.97	0.58	2.04	0.99 120
Variance of notmaly distributed *stances in a random populations, a $\sigma r^2 = 0.06123/p$	193.48	170.59	93.69	33.65	416.46	143.79
Variance of corrected mean nearest neighbour distances $(\sigma r2)\left[\sigma r2 = \sigma r^2 + 0.37\sqrt{(1/\rho)}^* (L/n)\right]$	235.47	4217	113.83	4125	500.02	174.33
Standard arse of the expected mean nearest neighbour distance sr = $\sqrt{(\sigma r^2/n)}$	1.39	1.31	0.97	0.58	2.04	1.20
Standard vaiate of the normal curve (c) = (rA - rE)/(rE)	-10.02	4.10	-14.68	-7.47	-3.48	-0.96

The same was observed for sample plots in Mwingi District. However, trees on farms in Mbeere District were significantly more dispersed than those in bushes. Pair wise comparisons of the mean nearest neighbour distances between individual study plots using the Mann Whitney test showed all were significantly different (Z values -2.649 to -9.213, p values 0.01 to < 0.001) except between those in Mwingi; those in Kitui and the ones in bushes in Mwingi and Mbeere Districts.

Tree Morphological Characteristics

Tree Height

Tree heights were not significantly different between the two land use practices across districts. Over 40% of trees from Kitui District were in the ≤ 5 m and 5 – 7 m height classes each, but 40% of those in Mbeere District were in the 5 – 7 m and 30% in the 7 – 9 m classes (Figure 2). Trees from Mwingi District were mainly in the intermediate height classes between these extremes. Mean heights differed significantly between individual study plots (Table 2). The trees in Kitui District were significantly shorter while trees in Mbeere were the tallest. In all districts, the trees on farms were taller than those in bushes (Table 2). Analysis of variance revealed that the tree heights in the three districts were significantly different from each other at 95% confidence interval (Table 2).

Diameter at Breast Height (DBH)

V. payos populations sampled were dominated by trees within the range of 10 to 20 cm diameter with over 55% of the trees (Figure 3). In Kitui and Mwingi Districts, the 10 - 20 cm diameter class contributed 60 and 76% of their population on farms and bushes, respectively. Trees with diameters above 30 cm made up to 8 and 7% of the trees in the two districts respectively. There was no clear difference in terms of tree diameters on farms and in bushes in the two districts. Trees in Mbeere in the 10 - 20 cm diameter class accounted for 25% of the population while trees with diameters higher than 30 cm were more on farms (36%) as compared to bushland (20%). Diameters differed significantly among plots in Mbeere with trees from this district differing significantly from those of the other two districts at $p = 0.05$ (Table 2). Mean diameters were higher in *V. payos* trees found in bushlands in Mwingi and Kitui Districts while the trees found on the farms in Mbeere had higher diameters than in those that were in the bushlands.

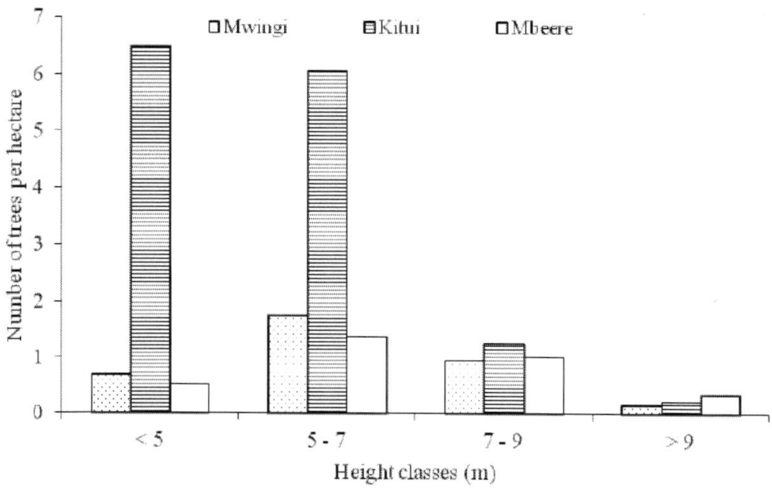

Figure 2: Height class distribution of *V. payos* trees in Mbeere, Mwingi and Kitui Districts in Eastern province, Kenya.

Table 2: Mean *Vitex payos* tree heights, dbh and crown diameter in three districts in Eastern province, Kenya

District	Height mean (m)	Mean dbh (cm)	Mean crown diameter (m)
KU	5.25±0.10a	18.0 ± 0.4a	4.86±0.10a
Mwingi	6.16 ± 0.11b	18.1 ± 0.5a	5.50 ± 0.12b
Mbeere	6.80±0.13c	24.4 ± 0.6b	5.81 ± 0.13b

Tree Crown Diameter

Over 77% of trees in all populations were in 3 – 5 m and 5 – 7 m crown diameter classes. Among trees in bushland of Kitui District,

67% of the tree population had crown diameters of up to 5 m while those in cultivated sites had 51% of their population with crown diameters in the same range. The two study plots had 8% of the trees with crown diameters of more than 7 m. In Mwingi District, 38% of the trees had crowns with diameter of up to 5 m while 17% of the tree population had above 7 m crown diameters. There were more trees in the bushes with high crown diameters at 20% than on farms (15%). In Mbeere District, there were more trees with crown diameters less than 5 m (41%) and above 7 m (26%) on the farms than in the bushes (23%). Analysis of variance revealed significant differences between tree crown diameters in districts (Table 2). There were no significant differences between tree crown diameters of trees on farms and in the bushlands. However, the trees on farms had slightly bigger crown diameters in Mwingi and Kitui Districts. Mbeere trees in bushes had bigger crowns than those on farms though the difference was minimal.

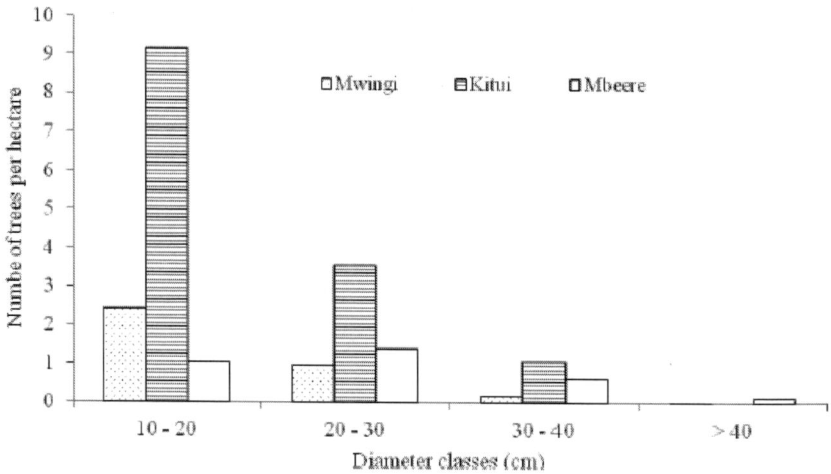

Figure 3: Diameter class distribution of *V. payos* trees in Mbeere, Mwingi and Kitui Districts, Kenya.

Table 3: Mean number of *V. payos* seedlings per hectare from different land uses three districts in Eastern province, Kenya

District	Land use		Origin of seedlings		
		Seed	Root	Stump/ coppice	Combined
Mbeere	On farm	0.04a	0.04a	5.02a	6.00a
Mwingi	On tam	1.45a	6.55bc	2.91a	10.91ab
Mwingi	Bushland	4.08ab	1.06ab	1.06a	8.00ab
Kitui	Bushland	14.67bc	38.6d	47.33c	100.7c
Kitui	On farm	17.33bc	4.44bc	7.56ab	29.03b
Mbeere	Bushland	47.56c	16.44cd	13.78b	77.78c

Mean ± (sem) number of seedlings followed by same letter are not significantly different at $p = 0.05$.

Regeneration of *V. payos*

Seedling Populations

The highest numbers of seedlings from all origins were recorded in bushlands in Kitui and Mbeere Districts with combined 101 and 78 seedlings per hectare, respectively. The farm site in Kitui had 29 seedlings per hectare. Other study sites had fewer seedlings, with less than 10 individuals from all three sources, that is, seed, stumps and root sprouts. Sixty one percent of seedlings in bushlands in Mbeere were mainly from seed while 46.7% of those in bushlands in Kitui were from stumps and 38.5% from root suckers. The farm site in Kitui had a moderate population of seedlings (17 seedlings per hectare) of seed origin, which were more than those in bushlands (15 seedlings per hectare). On the farms in Mwingi, there were seedlings from root suckers that were protected by farmers. There were seven seedlings per hectare from stumps on farms compared to two seedlings from seed origin per hectare. In bushlands in Mwingi, regeneration was very low as was the case on farm in Mbeere District. In these study plots, regeneration was very low irrespective of the seedling origin with six and eight seedlings per hectare respectively (Table 3). Analyses of variance revealed that there were significant differences between seedling numbers from individual origins and combined origins between the sample plots (Table 3). In the Mwingi bushlands, there were high regeneration from seed while in Kitui

bushlands, there were more seedlings regenerated from the root sprouts and coppices (Table 3). There were significant differences in seedling population from all origins among the three districts. Among seedlings of seed origin, Mwingi had the lowest population (6 seedlings per hectare) while Mbeere had the highest (48 seedlings per hectare) (Table 4). However, among seedlings from root sprouts and coppices, Mwingi had the lowest population (8 and 4 seedlings per hectare) while Kitui District had the highest number (43 and 55 seedlings per hectare respectively). Overall Kitui had highest seedling population though not significantly different from those in Mbeere. Combining all the districts, seedlings from all origins had significantly higher densities in bushes (55.36±11.35) than on farms (14.8±3.92).

Table 4: Mean number of *V. payos* seedlings per hectare from Mwingi, Mbeere and Kitui districts in Eastern province, Kenya

		Origin of seedlings		
District/land use	Seed	Root	Stump/ coppice	Combined
Mwingi	3.05 ± 1.68a	4.19 ± 2.15a	2.29 ± 0.76a	9.52 ± 2.81a
Mbeere	22.74 ± 8.37b	8.00 ± 3.28ab	9.26 ± 2.27b	40.00±11.22b
Kitui	16.27 ± 5.07b	18.13 ± 6.33b	23.47 ± 9.13b	57.87±14.63b

Mean ± (sem) number of seedlings followed by same letter are not significantly different at p = 0.05.

Table 5: Mean number of saplings per hectare from different origins in three districts in Eastern province, Kenya (± sem)

District	Land use		Origin of saplings		
		Seed	Root	Stump/ coppice	Combined
Kittui-	On farm	0	3.56 ± 2.05	7.56 ± 5.23	11.11 ± 6.79
Mbeeree	On farm	0	0	1.2 ± 0.85	1.2 ± 0.85
Mwingi	On farm	0.36 ± 0.36	1.09 ± 1.09	2.18 ± 1.46	3.64 ± 2.50
Mbeere	Bushland	0.44 ± 0.44	1.78 ± 1.35	4.44 ± 2.62	6.67 ± 3.64
Mwingi	Bushland	2.8 ± 1.34	1.2 ± 0.85	2 ± 1.37	6 ± 2.40
Kittui	Bushland	3.33 ± 2.17	19.33 ± 12.33	12.67 ± 4.43	35.33 ± 18.17
p value		0.055	0.145	0.163	0.216

Sapling Populations

Sapling population densities were low, ranging from one sapling per hectare on farms in Mbeere District to 35 saplings per hectare in bushlands in Kitui. The saplings were generally few on farms with an overall mean of five saplings against a mean of 16 saplings in bushlands from the three study districts. In all locations, saplings of coppice origin dominated except in bushlands in Kitui where those from root suckers were dominant and in bushlands in Mbeere where saplings of seed origin were dominant (Table 5). On farms in Kitui and Mwingi Districts, there were no saplings of seed origin encountered, while on farm in Mbeere, only saplings of coppice were encountered. Analyses revealed no significant differences between saplings from all origins between the study sites (Table 5). Between the land uses, mean number of saplings of seed origin differed significantly (p = 0.009) while those from root suckers, coppices and all sources combined were not statistically different. Although there were no significant difference of sapling from root sprouts and stump coppices between on farm and bushlands, the number was higher in the bushlands in all cases. Comparison of the combined population of seedlings in the two land uses, saplings and trees, revealed a normal population trend though weak in bushlands (Figure 4). On cultivated sites, a reverse trend was observed: on farms in Mwingi District, there were more saplings than seedlings; while in Kitui and Mbeere Districts study sites, there were more trees than saplings (Figure 4).

DISCUSSION

Population Structure

V. payos trees on farms were normally clustered in aggregates based on the farmers' acceptance of the species, the past land use practices and the current level of utilization of the trees and their products. Some farmers have left some trees in clusters on the farm for commercial harvesting of the fruits while others spared individual trees for provision of subsistence fruits to their households. Other trees were retained to provide sites for beehive keeping on the farms or shade for animals during dry season. Tree population densities expressed on a per hectare basis showed that there were fewer trees on farms than in bushes in all districts. In Mbeere, where the fruits are widely consumed with no sales, there were correspondingly fewer trees per hectare on farms than in bushlands. While in Mwingi District, where some farmers were involved in sale of fruits, the density of *V. payos* on the farm and in bushes were more or less equal. Farmers in this district were divided on the actual position of these fruits as commodity of trade. While some farmers left their fruits to waste on the ground, a few collected and took to markets. As a result, the maintenance and appreciation of the species was moderate, thus the low density of trees. In Kitui, the density of *V. payos* trees was higher than the other places; and higher in bushlands than on farms. Kitui study area was situated between two local markets, Ikanga and Mutomo, where these fruits are sold when in season. Thus traders frequented the area and many farmers were involved in on-farm sale of *V. payos* fruits.

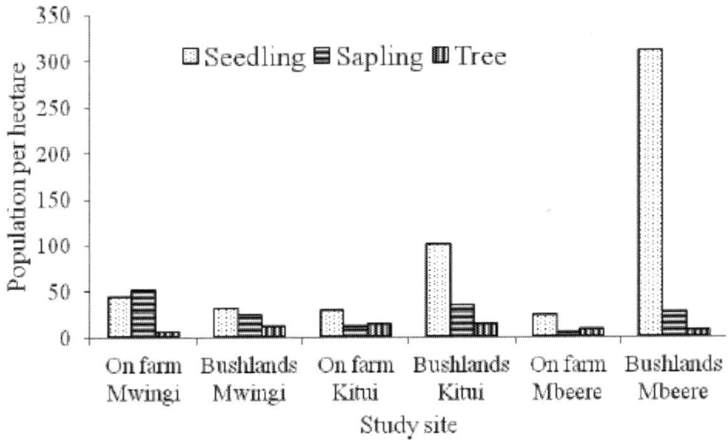

Figure 4: *V. payos* seedlings,saplings and trees in three districts in Eastern province, Kenya.

The spatial distribution of the species did show aggregated clumps of trees scattered on farms. This indicated selective removal of trees and/or poor regeneration of the species. Among the six study populations, trees aggregation was high in Kitui and on farm sites in Mwingi with mean distances between nearest neighbouring trees being less than 15 m. The cause of this situation was retention of clumps of trees by individual farmers, rather than all sampled trees being close together on farms in the two districts. *V. payos* trees in Kitui District bushlands were scattered closely together as a stand in woodland whose farming activities were minimal and the owners were of advanced age. This clumped spatial distribution may be related to the past cutting for firewood and charcoal and the extensive coppicing and suckering ability of species. In the villages with high mean distances between trees, several conserved trees were isolated on farms away from their nearest neighbouring trees (e.g. Mbeere District bushlands). While He et al. (1997) pointed out that the spatial distribution of a population shifts from highly aggregated to more random or uniform with succession, the tree population in the study areas was very low. Further, both abiotic (e.g. drought) and biotic factors (e.g. grazing, cutting, poor germination) may have greatly contributed to the current distribution pattern. According to Jama et al. (2008), the density of indigenous fruit trees often decreases with increasing human population pressure. Tree distribution in Mbeere was sparser than in the other two districts. The use level of the *V. payos* trees in Mbeere was low with only a few trees retained by individual farmers for provision of subsistence fruits and setting of beehives. Their crowns were said to be incompatible with both tobacco and 'khat' (*Catha edulis*) crops on farms, two cash crops extensively cultivated in Mbeere study area. The height structure of *V. payos* showed that there is a consistent size class distribution trend in Mbeere and Mwingi Districts, which differed from that of Kitui. Stands in Kitui were predominantly in an area that experienced less precipitation than the other areas. The area formed the extreme edge of natural range of species to the south. It was therefore possible that the trees were not in their optimal conditions in terms of moisture, and thus the stunted growth. The diameter growth of the trees also followed the same trend, with tree sizes increasing as the environmental conditions (rainfall) improves from Kitui through Mwingi to Mbeere. Seedlings and saplings are by virtue of their size, greatly influenced by intensity, magnitude and frequency of disturbance they are subjected to (Khan et al., 1987; Duchok et al., 2005). In the present study, destruction of seedlings occurred mainly in the cultivated farmlands partly due to little appreciation of these trees by the farmers. During land preparation and weeding of crops, the seedlings and saplings of *V. payos* were normally uprooted together with other weeds. In the field, seedlings originating from seed could fail to grow due to soil compaction or destruction of seed before germination. Preliminary observations of the fruits of *V. payos* revealed that most of them contained few viable seed, either due to poor pollination or to fungal infection of embryo immediately after pollination (Kimondo, 2010). Duchok et al. (2005) showed that germinated seedlings of *Illicium griffithii* were mainly found below fruiting trees. *V. payos* seedlings germinated below canopies of fruiting trees in bushes and cultivated areas. Nevertheless, in the bushes, these sites were normally the resting places for animals in the field especially during the dry hot season. Animals trampled upon the young seedlings causing heavy mortality. Kindeya (2003) quoted in Ogbazghi et al. (2006) reported that the seedlings in grazing fields often suffer as high as 90% mortality. Young seedlings were also prone to death due to effects of drought. In the present study, saplings on farms were frequently cut back to reduce the area they occupied thereby curtailing their development into mature trees. During the dry season, green foliage of *V. payos* saplings provided material for covering traditional earth kilns for charcoal production (Kimondo, 2010). In the fields, livestock also heavily browsed on crown of saplings while others were broken. Transition of saplings into trees was therefore very low. This seems to imply that there was generally poor natural regeneration status of *V. payos* in all the study areas, though the situation was more critical on the farms. Variation of trees in size class distribution in natural stand should ideally follow a normal distribution. Many factors could cause some deviations in different locations. Ogbazghi et al. (2006) indicated that the deviation could be caused by the pattern of regeneration and anthropogenic factors assuming local environmental conditions are the same. In Mwingi, it was noted that *V. payos* trees compete for space on the farms with several introduced fruits such as citrus, mangoes, avocados and guava among others. The exotic fruits have ready market locally, and they fetch higher prices than the indigenous fruits. As such, indigenous fruit trees were cut to create space for agricultural crops and exotic fruit trees. The occurrence of burnt stumps of indigenous trees, including *V. payos* on farms near traditional earth charcoal kilns indicated the aim was to obtain charcoal and suppress the tree coppicing. There was marginal utilization of *V. payos* fruits in Mbeere mainly by children. The high income realized from both 'khat' and tobacco necessitated cutting of most *V. payos* and other indigenous trees to minimize competition of resources such as light, water and nutrients. High germination rate of *V. payos* seed in bushland in Mbeere occurred on land whose owners were absent (Kimondo, 2010). Fallen ripe fruits generally remained on the ground with minimal collection. After rains, these germinated directly below the tree. However, during the dry season, animals browsed and trampled the young seedlings resulting in minimal translation into saplings. In a stable tree population, the number of seedlings should be higher than that of saplings and saplings numbers higher than that of trees (Duchok et al., 2005). The situation in Mbeere and Mwingi may without any efforts to plant or conserve the regenerates of *V. payos*, lead to the species becoming rare. The fact that most seedlings and saplings originated from root suckers and coppices is not healthy for the survival of *V. payos* in the study areas. It was therefore concluded that in such a situation, future planting stocks could be clones of existing trees as few seedlings from the seeds do mature into productive trees (Neke et al., 2006). The benefits associated with new genetic combinations shall thus be limited where regeneration through seed remains suppressed. This could also eventually result in regenerating trees close to parent trees, thus hindering exploitation of potentially new niches. Crown diameter of *V. payos* ranged from 3 to 9 m though trees with greater diameters were observed especially on cultivated or fallow sites. This implied that with appropriate tending of trees and reduction of competition, it is possible to increase the crown volume and therefore fruit production. The large sized tree crowns in Mbeere could have been influenced by the higher rainfall received in the area. However, few big sized trees with large crowns still exist in areas experiencing harsh environmental conditions and sandy soils in Kitui and Mwingi. The production of fruits depends on an extensive crown of a vigorously growing tree. Identification of such trees growing in extreme marginal areas may provide future well adapted germplasm for planting.

CONCLUSIONS AND RECOMMENDATION

This study was aimed at investigating the population structure and regeneration status of *V. payos* in the drylands of Kenya. The spatial distribution pattern of mature *V. payos* trees showed an aggregation of trees both on farms and in bushes. Tree densities ranged from 1.6 trees per hectare on farmlands to 20.3 trees in the bushes. The expected mean distances between nearest neighbouring trees were higher than the observed values in all sites, confirming that *V. payos* trees were more aggregated than randomly dispersed. The mean heights of the trees varied from 5.19 to 7.05 m cross the three study districts. Tree heights in the three districts were significantly different from each other. The sampled populations were dominated by trees (55%) with a mean diameter ranging from 18.0 to 24.4 cm and mean crown diameter ranging from 4.86 to 5.8 m. Highest densities of seedlings from all origins (seed, stumps and root sprouts) were recorded in the bushes in Kitui and Mbeere Districts with combined 101 and 78 seedlings per hectare respectively. Regeneration of seedlings was very low in Mwingi and some areas in Mbeere. Sapling densities were generally low with an overall mean of five saplings per hectare on the farmland against a mean of 16 per hectare in bushes from the three study areas. *V. payos* fruits are generally valued in some areas as a subsistence and commercial product. However, because of the high preferences of exotic fruits coupled with an increasing human population in the study areas, it was observed *V. payos* among other indigenous fruit trees continue to be cut down. Furthermore, natural regeneration was affected by livestock through browsing and trampling. Considering the adaptability and resilience of *V. payos* in these harsh environmental conditions in drylands, there is a need to promote its wide scale utilization and conservation. Superior trees in terms of tree size and the desired fruit traits should be sought to provide planting germplasm.

REFERENCES

1. Campbell B (1996). Aggregation and regularity: An inclusive one tailed nearest neighbour analysis of small spatially patchy populations. Oecologia 106:206-211.
2. Clark PJ, Evans FC (1954). Distance to nearest neighbour as a measure of spatial relationships in populations. Ecology 35(4):445- 453
3. Dancey CP, Reidy J (2007). Statistics without maths for Psychology. Prentice Hall. 619 pages.
4. Donnelly K (1978). Simulations to determine the variance and edgeeffect of total nearest neighbour distance. In: Hodder, I. (ed.) Simulation methods in archaeology. Cambridge University Press, London, pp. 91-95.
5. Duchok R, Kent K, Khumbongmayum AD, Paul A, Khan ML (2005). Population structure and regeneration status of medicinal tree *Illicium griffithii* in relation to disturbance gradients in temperate broad-leaved forest of Arunachal Pradesh. Curr. Sci. 89(4):673-676.
6. Dunn JE (1991). Why can't we see the indigenous wood for the exotic Trees? Glo. Ecol. Biogeogr. Lett. 1(2):33-35.
7. Ekesa BN, Walingo MK, Abukutsa-Onyango MO (2009). Accessibility to and consumption of indigenous vegetables and fruits by rural households in Matungu division, Western Kenya. Afr. J. Food Agric. Dev. 9(8):1725-1738.

8. He FL, Legendre P, LaFrankie JV (1997). Distribution patterns of tree species in a Malaysian tropical rain forest. J. Veg. Sci. 8:105-114.

9. Jama BA, Mohamed AM, Mulatya J, Njui AN (2008). Comparing the Big Five": A framework for the sustainable management of indigenous fruit trees in the drylands of East and Central Africa. Ecol. Indicat. 8:170-179.

10. Kauter D, Lewandowski I, Claupeina W (2003). Quantity and quality of harvestable biomass from Populus short rotation coppice for solid fuel use—a review of the physiological basis and management influences. Biomass Bioenergy 24:411-427.

11. Khan ML, Rai JPN, Tripathi RS (1987). Population structure of some tree species in disturbed and protected sub-tropical forests of northeast India. Acta Ecol. *8:*247–255.

12. Kimondo JM (2010). The potential for optimization of *V. payos* as a dryland resource in Kenya. PhD thesis, University of Wales, Bangor UK.

13. Kindeya G (2003). Ecology and management of *Boswellia papyrifera* (Del.) Hochst. dry forests in Tigray, Northern Ethiopia. PhD thesis Georg-August-University of Göttingen, Germany.

14. Kleinbaum DG, Kupper LL (1978). Applied regression analysis and other multivariate methods. Duxbury Press, North Scituate, Massachusetts. 556pp.

15. Luoga EJ, Witkowski ETF, Balkwill K (2004). Regeneration by coppicing (resprouting) of miombo (African savanna) trees in relation to land use. For. Ecol. Manage. 189:23-35.

16. Maundu PM, Ngugi GW, Kabuye CHS (1999). Traditional food plants of Kenya. English Press Ltd. Nairobi, Kenya. National Museums of Kenya.

17. Mbora A, Jamnadass R, Lillesø JPB (2008). Growing high priority fruits and nuts in Kenya: Uses and management. Nairobi. The World Agroforestry Centre. 61 pages

18. Morris RJ (2010). Anthropogenic impacts on tropical forest biodiversity: a network structure and ecosystem functioning perspective. Phil. Trans. R. Soc. B 365:3709–3718.

19. Neke KS, Owen-Smith N, Witkowski ETF (2006). Comparative resprouting response of savanna woody plant species following

harvesting: the value of persistence. For. Ecol. Manage. 232:114-123.

20. Ogbazghi W, Bongers F, Rijkers T, Wessel M (2006). Population structure and morphology of frankincense tree Boswellia papyrifera along an altitude gradient in Eriteria. J. Drylands 1(1):85-94.
21. Oroda ASK (2011). The impact of increased human population and sedentarisation of the pastoral communities on the land cover and the resources of Mt. Marsabit forest and the surrounding lands. MSc Thesis, Kenyatta University, 135 pages.
22. Sparks JC, Masters RE, Payton ME (2002). Comparative evaluation of accuracy and efficiency of six forest sampling methods. Proc. Okla. Acad. Sci. 82:49-56.
23. Turner RM, Ochung HA, Turner JB (1998). Kenya's Changing Landscape. The University of Arizona Press. Tucson. J. Arid Environ. 40(4):341-345.
24. Wanyonyi RW (2012). Impact of human population on land degradation in former Lugari District, Kakamega County, Kenya. MSc Thesis, Kenyatta University. 137 pages.

Chapter 6

Carbon Sequestration Potential of Agroforestry Systems in India

Indu K Murthy, Mohini Gupta, Sonam Tomar, Madhushree Munsi, Rakesh Tiwari, GT Hegde, and Ravindranath NH

Centre for Sustainable Technologies, Indian Institute of Science, Bangalore, India

ABSTRACT

Forestry has been recognized as a means to reduce CO_2 emissions as well as enhancing carbon sinks. Forests are a large sink of carbon and their role in carbon cycles is well recognized. This paper reviews the role of agroforestry systems in carbon mitigation. Agroforestry provides a unique opportunity to combine the twin objectives of climate change adaptation and mitigation. It has the ability to enhance the resilience of the system for coping with the adverse impacts of climate change.

Agroforestry systems offer important opportunities for creating synergies between both adaptation and mitigation actions. Various authors have carried out studies to estimate carbon stocks in different agroforestry systems in India. Agroforestry systems have the potential to provide significant mitigation options but they require proper management that influences the amount of carbon sequestered. The role of agroforestry practices in climate change mitigation in India can be realized to its full potential by overcoming various technical, financial and institutional barriers.

INTRODUCTION

There is a growing interest in the role of different types of land use systems in stabilizing the atmospheric CO_2 concentration and reducing the CO_2 emissions or on increasing the carbon sink of forestry and agroforestry systems. Forestry has been recognized as a means to reduce CO_2 emissions as well as enhancing carbon sinks. The role of forests (or trees) in carbon cycles is well recognized [1] and forests are a large sink of carbon [2,3]. There is considerable interest to increase the carbon storage capacity of terrestrial vegetation through land-use practices such as afforestation, reforestation, and natural regeneration of forests, silvicultural systems and agroforestry [4,5]. Agroforestry systems are very important given the area currently under agriculture, the number of people who depend on land for their livelihoods, and the need for integrating food production with environmental services [6-8].

Globally, climate negotiations have highlighted the importance of land use sectors in mitigating the climate change. Agriculture alone accounts for 10-12% of the total global anthropogenic emissions of GHGs with an estimated non-CO_2 GHG emission of 5120-6116 $MtCO_2$ eq/yr in 2005 [9]. Since agricultural lands are often intensively managed, they offer many opportunities to improve agronomic practices, nutrient and water management, land use practices to fit the land managers' objectives of carbon sequestration. The total carbon sequestration potential of global croplands is about 0.75-1Pg/yr or about 50% of the 1.6-1.8 Pg/yr lost due to deforestation and other agricultural activities [10].

The emphasis of land use systems that have higher carbon content than existing plant community can help achieve net gains in carbon, specifically and significant increases in carbon storage can be achieved by moving from lower biomass land uses [e.g. grasslands, crop fallows, etc] to tree based systems such as forests, plantation forests and agroforestry [11]. Agroforestry provides a unique opportunity to combine the twin objectives of climate change adaptation and mitigation. Although agroforestry systems are not primarily designed for carbon sequestration, there are many recent studies that substantiate the evidence that agroforestry systems can play a major role in storing carbon in aboveground biomass [12-16] and in soil [17] and in belowground biomass [17]. Some of the earliest assessments of national and global terrestrial carbon dioxide sinks reveal two beneficial attributes of agroforestry systems (a) direct near term storage [decades to centuries] in trees and soils and (b) the potential to offset immediate GHG emissions associated with deforestation and subsequent shifting cultivation [18].

In this paper, we present a review of carbon sequestration potential of agroforestry systems, particularly India and highlight the need for more studies estimating the potential of agroforestry systems given the multiple benefits of such systems and the potential to provide synergy between climate change mitigation and adaptation by way of decreasing the vulnerability of communities to climate risks and climate change in the long run.

CARBON SEQUESTRATION POTENTIAL OF AGROFORESTRY SYSTEMS

Agroforestry, the practice of introducing trees in farming has played a significant role in enhancing land productivity and improving livelihoods in both developed and developing countries. Although carbon sequestration through afforestation and reforestation of degraded natural forests has long been considered useful in climate change mitigation, agroforestry offers some distinct advantages. The planting of trees along with crops improves soil fertility, controls and prevents soil erosion, controls water logging, checks acidification and eutrophication of streams and rivers, increases local biodiversity, decreases pressure on natural forests for fuel and provides fodder for livestock [8]. It also has the ability to enhance the resilience of the system for coping with the adverse impacts of climate change.

The effectiveness of agroforestry systems in storing carbon depends on both environmental and socio-economic factors; in humid tropics, agroforestry systems have the potential to sequester over 70Mg/ha in the top 20 cm of the soil [14].The carbon storage capacity in agroforestry varies across species and geography [19]. Further, the amount of carbon in any agroforestry system depends on the structure and function of different components within the systems put into practice [20,21].

The fact that agroforestry systems can function as both source and sink of carbon has been presented in literature [13,18]. There is also clear evidence to suggest that the type of agroforestry system greatly influences the source or sink role of the trees. For example, agrisilvicultural systems where trees and crops are grown together are net sinks while agro silvipastoral systems are possibly sources of GHGs [22]. Practices like tillage, controlled burning, manuring, application of chemical fertilizers and frequent soil disturbance can lead to significant emissions of GHGs [13]. According to the IPCC [9] agroforestry systems offer important opportunities of creating synergies between both adaptation and mitigation actions with a technical mitigation potential of 1.1-2.2 PgC in terrestrial ecosystems over the next 50 years. Additionally, 630 Mha of unproductive croplands and grasslands could be converted to agroforestry representing a carbon sequestration potential of 391,000 MgC/yr by 2010 and 586,000 MgC/yr by 2040 [23]. The carbon in the aboveground and belowground biomass in an agroforestry system is generally much higher than the equivalent land use without trees (i.e. crop land without any trees).

The estimates of potential for carbon storage in different kinds of agroforestry systems are provided in Table 1. In Southeast Asia, agrisilvicultural systems have the capacity to store 12-228 MgC/ha in humid tropical lands and 68-81 MgC/ha in dry lowlands. Highest potential for carbon storage can be observed for North American silvi pastoral systems with a range of 90-198 MgC/ha. The potential to sequester carbon in aboveground components in agroforestry systems is estimated to be 2.1×10^9 $MgCyear^1$ in tropical and 1.9×10^9 $MgCyear^1$ in temperate biomes [24].Agroforestry systems can have indirect effects on carbon sequestration as it helps decrease pressure on natural forests that are the largest sinks of terrestrial carbon, they also conserve soils and thus enhance carbon storage in trees and soils. Effects of agroforestry practices on the soil carbon pool indicated a rate of increase by 2-3 MgC/ha/yr [25]. Estimations of carbon sequestration potential in various studies report an estimated potential of 6.3GtC [26] and 0.7-1.6 GtC [27].

The carbon sequestration potential of agroforestry systems has been established theoretically; however field measurements to validate these concepts are limited. The inherent variability in the estimates of potential carbon storage in agroforestry systems and the lack of uniform methodologies has made comparisons difficult [23]. Few studies of specific agroforestry practices have proved potential for carbon sequestration [28,29].

Table 1: Carbon storage potential of agroforestry systems in different eco-regions of the world (80)

Continent	Eco region	System	Potential (Mg Cha-1)
Africa	Humid tropical high	Agrosilvicultural	29-53
South America	Humid tropical low dry lowlands		39-102 39-195
Southeast Asia	Humid tropical dry lowlands		12-228 68-81
Australia	Humid tropical low	Silvipastoral	28-51
North America	Humid tropical high humid tropical low dry lowlands		133-154 104-198 90-175
Northern Asia	Humid tropical low		15-18

AGROFORESTRY IN INDIA

India has a long tradition of agroforestry practices. The agroforestry systems in India include trees on farms, community forestry and a variety of local forest management and ethno forestry practices [30]. In India, the practice of growing scattered trees on farmlands is quite old and has not changed much over centuries; these trees are multipurpose, used for shade, fodder, fuel wood, fruit, vegetables and medicinal uses. Trees like *Eucalyptus* and *Populus* are also grown in agricultural fields or on field bunds often on farm boundaries in Punjab and Haryana. Shifting cultivation in the Northeast India and Taungya cultivation in Kerala, West Bengal, and Uttar Pradesh and to a limited extent in Tamil Nadu, Andhra Pradesh, Orissa, Karnataka, as well as in the Northeast hill regions are examples of traditional Indian agroforestry systems. Besides, home gardens, wood lots, large cardamom plantations of Eastern Himalayas and plantations elsewhere and Alder based agriculture in Northeast India are other kinds of agroforestry systems.

GEOGRAPHICAL SPREAD OF TRADITIONAL AGROFORESTRY SYSTEMS

Shifting cultivation, homegardens and plantation-based cropping systems are mostly practiced in humid tropical regions. In India, home gardens are found in Kerala and Andaman and Nicobar Islands. Taungya, boundary plantations, live hedges, range land trees are agroecologically adapted to all regions. Boundary plantations are found in Uttar Pradesh, Gujarat, in parts of south India, particularly the Nilgiri hills, Haryana, Himachal Pradesh, Bihar and Orissa. Woodlots are found in hilly areas whereas shelter belts are found in wind-prone regions like coastal areas. In India, woodlots are found in Andhra Pradesh, Tamil Nadu, Karnataka, Orissa, Punjab, Haryana, Gujarat and Assam. Scattered trees on farmlands are found in all regions specially arid and semi-arid. Agro-silvo-pastoral practices are found in semiarid regions of India. Below we describe the major agroforestry practices in four important regions of India.

Northeast India

The major form of agroforestry practiced in the region is shifting cultivation (jhum). The entire socio-economic structure is woven around this system and farmers maintain high species diversity. However with reduced shifting cultivation cycle, the system has become ecologically unsound and has resulted in forest degradation. Agroforestry has the potential of restoration and maintenance of soil fertility, and increase in productivity. Some of the agroforestry systems followed in the area are Agri-horticulture, Silvipastoral, Agri-silviculture, Silvi-horticulture, Pastoral-silviculture and home gardens [31-35].

Northwest India

The states of Gujarat and Rajasthan receive scanty and erratic rainfall and failure of crop is a regular phenomenon. Agroforestry practices have the potential to minimize the impact of extreme climate condition and can provide alternate income tothe people. One of the major practices involves combining *Prosopis cineraria* with other agricultural crops in this region [36,37].

The Western Ghats

Managed ecosystems of Western Ghats present a complex pattern with a great diversity of trees and field crops. Agroforestry systems where trees are grown with crops, and/ or sometimes animals, in interacting combinations in space or time dimensions are abound in the Western Ghats region. In particular, plantation agriculture involving coffee (*Coffea spp.*), tea (*Camellia sinensis*), and spices in association with a wide spectrum of trees and para rubber (*Hevea brasiliensis*), rice-based cropping systems, coconut (*Cocos nucifera*) based cropping systems, and homestead farming systems dominate the region [38].

Southern India

In the southern states of Kerala, Tamil Nadu, Karnataka, and Andhra Pradesh, numerous forms of agroforestry are popular. Home gardens and multistory combinations involving plantation crops are prevalent in Kerala; tree-spice gardens, and crop combinations involving them are common in coastal Karnataka; energy plantations, especially of *Casuarina* spp., are popular in coastal districts of Tamil Nadu and Andhra Pradesh and various forms of intercropping with fruit trees and silk cotton tree (*Ceiba petandra*) are widespread in Tamil Nadu [39-42].

BENEFITS OF AGROFORESTRY SYSTEMS

Agroforestry has the potential to provide both economic and environmental benefits [20,43]. Some of the major benefits of maintaining agroforestry systems are discussed below.

Improved Soil Fertility

Enhancing and maintaining soil fertility is vital for food security, reducing poverty, preserving environment and for sustainability [30,44]. Agroforestry land use systems like agrohorticulture, agro-pastoral system, agri-silvipastoral system, etc., are efficient ways of restoring soil organic matter [30]. Studies have also shown that yield is higher with improved crop rotation than with continuous cropping [44]. The leaf litter from agroforestry practices, forms humus after decomposition and improves various soil properties [45]. Agroforestry can control runoff and soil erosion, thereby reducing losses of water, soil material, organic matter and nutrients. It can check development of soil toxicities, both soil acidification and salinization and trees can be employed in the reclamation of polluted soils.

Increased Income

The diverse component of agroforestry provides multiple harvests at different times of the year. It increases food production, improves supply of fodder for fish and livestock, increases supply of fuelwood, improves soil fertility and water supply, habitats, etc. Thus it reduces the risk of crop failure and ensures alternate income for the farmers [30].

Increased Carbon Stock

Agroforestry has a huge potential as mitigation strategy to the changing climate because of its potential to sequester carbon in its multiple plant species and soil [13,21,45]. The average carbon sequestered by these practices has been estimated to be 9, 21, 50, and 63 $MgCha^{-1}$ in semiarid, sub-humid, humid, and temperate regions. In tropics, for small agroforestry systems, it has been found to be ranging from 1.5 to 3.5 $MgCha^{-1}yr^{1}$ and thus can be a viable strategy for carbon storage [11,13]. In degraded soils of the subhumid tropics, agroforestry practices have been found to increase top soil carbon stocks up to 1.6MgCha-1yr^{1} [14]. Thus, proper designing and managing of agroforests can make them effective carbon sinks.

Reduced Vulnerability

Agroforestry increases the resilience of farming systems by buffering against various risks, both biophysically (hydraulic lift, soil fertility) and financially (diversification, income risk) [15]. Other advantages include reducing seasonal labor peaks, earn income throughout the year and ensure benefits over the short, medium and long term [46].

Increased Productivity

Studies show that forest influenced soils give higher yields than ordinary soils. Taungya cultivators got higher yields than from pure agriculture in Tarai region of Uttar Pradesh. IGFRI, Jhansi conducted experiments that indicated increased yield of fodder when fodder grasses were intercropped with fodder trees as compared to mono cropping of fodder grass. In South India, and states like Punjab, Haryana, Uttar Pradesh and Gujarat, intercropping agroforestry food crops was found to be more productive [47].

Aesthetic Value

Can create a healthy environment-interaction-Agroforestry from agroforestry practices can enhance the soil, water, air, animal and human resources of the farm. Agroforestry practices may use only 5%

of the farming land area, yet account for over 50% of the biodiversity, improving wildlife habitat and harboring birds and beneficial insects which feed on crop pests. Tree biodiversity adds variety to the landscape and improves aesthetics.

Other advantages include-utilization of solar energy more efficiently than monocultural systems, reduced insect pests and associated diseases, increased nitrogen inputs because of nitrogenfixing trees and shrubs and it can also moderate microclimates. Shelter given by trees improves yields of nearby crops and livestock.

CARBON STOCKS IN AGROFORESTRY SYSTEMS IN INDIA

Carbon sequestration in different agroforestry systems occurs both belowground, in the form of enhancement of soil carbon plus root biomass and aboveground as carbon stored in standing biomass. Some of the earliest studies of potential carbon storage in agroforestry systems and alternative land use systems for India had estimated a sequestration potential of 68-228 MgC/ha [2], 25tC/ha over 96 Mha of land [48]. But this value varies in different regions depending on the biomass production [30]. Studies done by Jha et al. [49] showed that agroforestry could store nearly 83.6 tC/ha up to 30 cm soil depth, 26% more carbon compared to cultivation in Haryana plains. However, the magnitude of carbon sequestration from forestry activities would depend on the scale of operation and the final use of wood.

AGRISILVICULTURAL SYSTEMS

Carbon Sequestration in Tree Biomass

Maikhuri et al. [50] estimated species wise annual carbon sequestration potential of planted tree species on abandoned agricultural land (3.9 t/ha/yr) and degraded forest land (1.79 t/ha/yr). The highest carbon sequestration was found for *Alnus nepaliensis* 0.256 tC/ha/yr and *Dalbergia sissoo* 0.141 tC/ha/ yr intercropped with wheat and paddy. In a 6 year old *Gmelina arborea* based agri-silvicultural system 31.37

tC/ha was sequestered [51]. In another study the carbon sequestration in monocropping of trees and food crops were 40% and 84% less than agri-silviculture indicating that agroforestry systems have more potential to sequester carbon [52]. In an agri-silvicultural system, *Dalbergia sissoo* at age 11 years was able to accumulate 48-52 t/ha of biomass [19]. Carbon dynamics involving different pruning treatments were studied in an agri-silvicultural system where tree biomass was 23.61 to 34.49 tC/ha with black gram-mustard. In a study on poplar based agri-silvicultural system, total biomass in the system was 25.2 t/ha, 113.6% higher than sole wheat cultivation, where net carbon storage was 34.61 tC/ha compared to 18.74 tC/ha in sole wheat cultivation [53]. In a system comprising *Albizia* and mixed tree species like Mandarin accumulated 1.3 t biomass/ha storing 6939kg/ha in tree and crop biomass was reported [54].

Soil Organic Carbon Enhancement

In a study conducted on intercropping of trees with crops, Singh et al. [55] reported an enhancement in SOC by 33.3 to 83.3% for *Populus deltoides* and *Eucalyptus* hybrid with *Cymbopogon* sp., with a greater increase in SOC under *Populus deltoides* plantation. Soil organic carbon has been reported to have improved for agroforestry plantations ranging in age from 6 years [50] to 20 years [56]. In a Poplar based agroforestry system, trees could sequester higher soil organic carbon up to 30cm depth during the first year of plantation (6.07 t/ha/yr) than in subsequent years (1.95-2.63 t/ha/yr) with greater soil carbon storage in sandy clays than loamy sand [57]. Traditional *Prosopis cineraria* based systems lead to a 50% increase in SOC largely due to leaf litter [58]. Samra and Singh [59] observed an increase in soil organic carbon status of surface soil by 0.39 to 0.52% under *Acacia nilotica+Sacchram munja* and 0.44 to 0.55% under *Acacia nilotica+Eulaliopsis binata* after 5 years.

SILVIPASTORAL SYSTEMS

Carbon Sequestration in Tree Biomass

Comparative studies conducted by NRCAF [60] on biomass production from natural grassland and silvipastoral system comprising *Albizia amara, Dichrostachys cinerea* and *Leucaena leucocephala* as woody perennials with *Chrysopogan fulvus* as grass and *Stylosanthes hamata* and *S. scabra* as legume revealed that in 8 years, rate of biomass carbon stored in silvipastoral system was 6.72 tC/ha/yr, two times more than 3.14 tC/ ha/yr from natural grassland. Singh [61] estimated the total carbon sequestered in farm forestry with species such as *Eucalyptus sp., Populus deltoides, Tectona grandis, Anthocephalus chinensis* trees to be around 16,400 t/yr. Rai et al. [62] studied the effect of introducing a silvipastoral system in a natural grassland in semi-arid Uttar Pradesh, where introduced species of *Albizia procera, Eucalyptus tereticornis, Albizia lebbeck, Embilica officinalis* and *Dalbergia sissoo* accumulated 8.6, 6.92, 6.52, 6.25 and 5.41 t/ha/yr of biomass. Here, the carbon storage in the system was 1.89-3.45 tC/ha in silvipasture and 3.94 tC/ ha in pure pasture. Saharan et al. [63] reported an increase in organic carbon of 1.7 to 2.3 times in a silvipastoral system involving *Leucaenea leucocephala, Cenchrus ciliaris* and *Stylosanthes hamata* compared to a control. Similar studies have reported higher organic carbon levels in dry sub-humid and arid ecosystems when grass species are intercropped with annual crops in a silvipastoral system with no increase in organic carbon with grasses in an arid ecosystem [58]. In a silvipastoral system, carbon flux in net primary productivity increased due to the integration of *Prosopis juliflora* and *Dalbergia sissoo* with grasses [64].

Table 2: Total biomass and carbon stock estimated for agroforestry systems in villages of Karnataka and Tamil Nadu, excluding the forest and plantation

Villages	Area under cropland (ha)	Total biomass* (tlha)	Total carbon stocks (tC/ha)	Total carbon stocks in villages (tClha)
Karnataka villages				
Sirsimakki	72.00	9.47	4.73	340.56
Hallusarige	198.80	1.62	0.81	161.03
Hegle	240.69	3.96	1.98	476.57
Lukkeri	58.00	8.51	4.25	246.50
Tamil Nadu villages				

Kempanaickena palayam	501.00	13.17	6.59	3301.59
Sellipalayam	389.00	10.15	5.07	1972.23
Thalakudi	77.00	3.51	1.76	135.52
Valadi	183.00	1.33	0.66	120.78
Total	1719.49			6754.77

*Includes both above and belowground biomass

Soil Organic Carbon Enhancement

Narain [65] reported that planting trees and grasses in degraded lands in arid areas can help increase soil carbon stock from 24.3Pg to 34.9 Pg. Studies conducted at RRS, Kukma have reported a total belowground carbon stock of 1.6 t/ ha, 23.4% of the total carbon stock under silvipastoral system involving *Acacia tortilis* and *Cenchrus setegerus* [66].

Carbon Stored in Block and Boundary Plantations

In a study done by Kumar [67] on four different agroforestry systems (*Populus deltoides* block plantation+wheat, *Eucalyptus* hybrid boundary plantation+wheat, *Populus deltoides* boundary plantation+wheat and *Populus deltoides* block plantation+lemon grass), all 9-year old block and boundary plantations, it was observed that total carbon sequestration [in trees] was 70.59, 21.38, 116.29 and 18.53 $tCha^{-1}$ in system *Populus deltoides*+wheat followed by 68.53, 20.63, 113.03 and 17.60 $tCha^{-1}$ in system *Populus deltoides*+Lemon grass, respectively, with a greater potential for carbon sequestration in boundary plantations of *Populus deltoides* and *Eucalyptus* hybrid. In a study on mitigation potential of block plantations in farm forestry, Hooda et al. [68] estimated a mitigation potential of 48.5, 62.7, 61.7, 60.8, 37.6 tC/ ha/yr respectively for Khair, Chir pine, mixed plantations, Mango and Kinoo based farm forestry systems in Uttaranchal.

BIOMASS AND CARBON STOCKS IN AGROFORESTRY SYSTEMS–CASE STUDIES FROM TAMILNADU AND KARNATAKA

A study was conducted by Indian Institute of Science as a part of the larger study on natural resource monitoring in selected village ecosystems of southern India falling in different agro-ecological zones. The study was conducted in 4 villages each in Karnataka and Tamil Nadu where all the trees on croplands were inventoried. Here we present the results of the agroforestry study conducted in the Karnataka villages that fall in the north Sahyadris and western Karnataka plateau and the central Karnataka plateau and the Tamil Nadu villages located in the Tamil Nadu uplands and plains agro-ecological zone [69].

All the villages of both Karnataka and Tamil Nadu practice agroforestry on irrigated as well as rainfed croplands–either as bund or block plantations. Coconut, followed by mango is the predominant species in agroforestry systems of the study villages. Other species include *Anacardium occidentale, Artocarpus integrifolia,* Areca and *Acacia auriculiformis* among others. The biomass and carbon stocks estimated in the agroforestry systems of the villages studied in Karnataka and Tamil Nadu are presented in Table 2.

Among the Karnataka villages, highest biomass carbon stock is recorded in Sirsimakki (about 5 tC/ha of biomass carbon), followed by Lukkeri, a coastal village (8.5 t/ha of biomass carbon). In the Tamil Nadu villages, the biomass stocks in 2 of the 4 villages is higher than that recorded in the Karnataka villages and it is in the range of 1.3 to 12 t/ha. The biomass carbon stock varies from village to village but is significant. Even if one takes an average of carbon stock (3.93 tC/ ha) and if one considers a conservative crop area of about 100 million hectare, the carbon stock in agroforestry systems could be estimated to be 392.84 MtC for India. This excludes soil organic carbon stocks.

CONCLUSIONS

Planting of multipurpose tree species in non-forest land categories serves a dual purpose, i.e. promotion of biodiversity and carbon sequestration. Trees serve as additional source of income at the time of crop failures [70,71]. They also provide economic rewards from the non carbon benefits [72,73]. A tree planting design with a combination of fast and slow growing species is useful, for different rotation rates of the two species render valuable wood that is economically beneficial, as well as wood for use as fuelwood and for construction purposes [74].

Further, trees on farms have the potential to improve productivity in two ways. Tree-crop combination can increase the amount of water that is used on farm and can also increase the productivity of water that is used by increasing biomass of trees or crops produced per unit of water. Evidence from semi-arid India and Kenya has shown that the greater productivity of agroforestry systems is primarily due to the higher amount of water used [75]. Higher biomass of ecosystems is associated with higher diversity, and higher species diversity leads to greater carbon sequestration. The management of agricultural lands will therefore play an important role in enhancing carbon sinks and in turn reducing emissions. Land use management measures such as conservation of existing tree cover, promotion of agroforestry, etc. will not only have positive impacts on biodiversity but also promote the use of biomass fuels, replacing the fossil fuels, thereby contributing to net reduction in CO^2 emissions. Also, an integrated farming approach would bring about change in soil quality, ground water level and thereby improving agricultural productivity, giving the farmer a diversity of products. Tree cultivation on agricultural land improves biomass productivity per unit area and also uses nutrients from different soil layers. Further, land such as bund and avenues that are hitherto not cultivated would increase the tree cover of the landscape [76].

Agroforestry systems show significant carbon accumulation in living biomass, as well as soil carbon, demonstrating the potential to offer the environmental service of carbon sequestration. Furthermore, agroforestry systems can contribute to reducing CO^2 emissions by avoiding burning of forest-based fuelwood and conserving soil. Besides the potential of agroforestry system to accumulate and sequester carbon, these systems could evolve into a technological alternative for reducing deforestation rates in tropical zones while also offering a wide variety of products and services to rural communities.

The Greening India mission under the National Climate Change Action Plan targets 1.5 Mha of degraded agricultural lands and fallows to be brought under agroforestry; about 0.8 Mha under improved agroforestry practices on existing lands and 0.7 Mha of additional lands under agroforestry [77].Much of the opportunity to store carbon through afforestation in India will occur through agroforestry on agricultural lands due to the fact that majority of arable land in India is being cultivated [78]. The total potential for agroforestry has been estimated at 25.36 Mha with almost half of it under tree borne oilseeds, silvipasture and others by 2025 [60].

The role of agroforestry practices in climate change mitigation in India can be realized to its full potential by overcoming technical barriers (management skills), financial and institutional (differing interests between farmers and forests departments and industry) and by implementation of targeted policies and incentives. There is need for research on the potential impacts of climate change on agroforestry systems. The development of appropriate silviculture and management techniques for shelterbelt and field bund planting requires careful attention in arid and semi-arid areas. Besides the potential of agroforestry system to accumulate and sequester carbon, these systems could evolve into a technological option for reducing the vulnerability of farming system to climate variability and climate change impacts. Agroforestry provides the best example of promoting mitigation and adaptation synergy in addressing climate change [79].

REFERENCES

1. *Singh R, Lal* M (2000) Sustainable forestry in India for carbon mitigation. Current Science 78: 563-567.
2. Dixon RK, Brown S, Houghton RA, Solomon AM, Trexler MC, et al. (1994) Carbon pools and fluxes of global forest ecosystems. Science 263: 185-190.
3. Wang X, Feng Z, Ouyang Z (2001) The impact of human disturbance on vegetative carbon storage in forest ecosystems in China. Forest Ecology and Management 148: 117-123.

4. Brown S (1996) Present and potential roles of forests in the global climate change debate. Unasylva 185: 3-10

5. Canadell JG, Raupach MR (2008) Managing forests for climate change mitigation. Science 320: 1456-1457.

6. Soto-Pinto L, Romero-Alvarado Y, Caballero-Nieto J, Segura WG (2001) Woody plant diversity and structure of shadegrown-coffee plantations in Northern Chiapas, Mexico. Rev Biol Trop 49: 977-987.

7. Garrity DP (2004) Agroforestry and the achievement of the millennium development goals. Agroforestry System 61: 5-17.

8. Makundi WR, Sathaye JA (2004) GHG mitigation potential and cost in tropical forestry-relative role for agroforestry. Environment, Development and Sustainability 6: 235-260.

9. Intergovernmental Panel on Climate Change, 2007. Assessment report.

10. Lal R, Bruce J (1999) The Potential of World Cropland to Sequester C and Mitigate the Greenhouse Effect. Environment Science and Policy 2: 177-185.

11. Roshetko JM, Lasco RD, Angeles MSD (2007) Small holder agroforestry systems for carbon storage. Mitigation and Adaptation Strategies for Global Change 12: 219-242.

12. Sathaye JA, Makundi WR, Andrasko K, Boer R, Ravindranath NH, et al. (2001) Carbon mitigation Potential And Costs Of Forestry Options In Brazil, China, India, Indonesia,Mexico, The Philippines and Tanzania. Mitigation and Adaptation Strategies for Global Change 6: 185-211.

13. Montagnini F, Nair PKR (2004) Carbon sequestration: An underexploited environmental benefit of agroforestry systems. Agroforestry Systems 61: 281- 295.

14. Mutuo PK, Cadisch G, Albrecht Palm CA, Verchot L (2005) Potential of agroforestry for carbon sequestration and mitigation of greenhouse gas emissions from soils in the tropics. Nutrient Cycling in Agroecosystems 71: 43-54.

15. Verchot LV, Noordwijk MV, Kandji S, Tomich T, Ong C, et al. (2007) Climate change: linking adaptation and mitigation through agroforestry. Mitigation and Adaptation Strategies for Global Change 12: 901-918.

16. Niles JO, Brown S, Pretty J, Ball AS, Fay J (2002) Potential carbon mitigation and income in developing countries from changes in use and management of agricultural and forest lands. Philosphical Transactions of The Royal Society Lond 360: 1621-1639.
17. Nair PKR, Nair VD, Kumar BM, Haile SG (2009) Soil carbon sequestration in tropical Agroforestry systems: a feasibility appraisal. Environmental Science and Policy 12: 1099-1111.
18. Dixon RK (1995) Agroforestry systems: sources or sinks of greenhouse gases? Agroforestry Systems 31: 99-116.
19. Newaj R, Dhyani SK (2008) Agroforestry for carbon sequestration: Scope and present status. Indian Journal of Agroforestry 10: 1-9.
20. Schroeder P (1993) Agroforestry systems: integrated land use to store and conserve carbon. Climate Research 3: 53-60.
21. Albrecht A, Kandji ST (2003) Carbon sequestration in tropical agroforestry systems. Agriculture, Ecosystems and Environment 99: 15-27.
22. Kandji ST, Verchot LV, Mackensen J, Boye A, Van NM, et al. (2006) Opportunities for linking climate change adaptation and mitigation through agroforestry systems: World Agroforestry into the Future, Garrity DP, Okono A, Grayson M, Parrott S: 113-121. World Agroforestry Centre (ICRAF), ISBN, Nairobi, Kenya.
23. Jose S (2009) Agroforestry for ecosystem services and environmental benefits: an overview. Agroforestry Systems 76: 1-10.
24. Oelbermann M, Voroney RP, Gordon AM (2004) Carbon sequestration in tropical and temperate agroforestry systems: a review with examples from Costa Rica and southern Canada Agriculture. Ecosystems and Environment 104: 359-377.
25. Garg VK (1998) Interaction of tree crops with a sodic soil environment: Potential for rehabilitation of degraded environments. Land Degradation and Development 9: 81-93.
26. Brown S, Sathaye J, Cannell M, Kauppi P (1996) Management of forests for mitigation of greenhouse gas emissions. In: Watson RT, Zinyowera MC, Moss RH (eds.), Climate Change 1995: Impacts, Adaptations and Mitigation of Climate Change: Scientific-Technical Analyses. Contribution of Working Group II to the Second Assessment Report of the Intergovernmental Panel on

Climate Change, Cambridge University Press, Cambridge and New York, USA.

27. Trexler MC, Haugen C, (1994) Keeping it Green: Tropical Forestry Opportunities for Mitigating Climate Change. World Resources Institute, Washington DC, PP. 52.
28. Roshetko JM, Delaney M, Hairiah K, Purnomosidhi P (2002) Carbon stocks in Indonesian homegarden systems: Can smallholder systems be targeted for increased carbon storage? American Journal of Alternative Agriculture 17: 138- 148.
29. Takimoto A, Nair PKR, Nair VD (2008) Carbon stock and sequestration potential of traditional and improved agroforestry systems in the West African Sahel. Agriculture, Ecosystems and Environment 125: 159-166.
30. Pandey DN (2007) Multifunctional agroforestry systems in India. Current Science 92: 455-463.
31. Ahmed AA, Hazarika DN (2007) Agroforestry systems and practices prevailing in Assam. In: Puri and Panwar, (eds.). Agroforestry systems and practices. New Delhi: New India Publishing Agency, India.
32. Tomar JMS, Upadhaya K, Tripathi OP, Pandey HN (2007) Agroforestry systems and practices prevailing in Meghalaya. In: Puri and Panwar, (eds.). Agroforestry systems and practices. New Delhi: New India Publishing Agency: 644.
33. Sahoo UK (2007) Agroforestry systems and practices prevailing in Meghalaya. In: Puri and Panwar, (eds.). Agroforestry systems and practices. New Delhi: New India Publishing Agency, India.
34. Sood KK, Handique P, Singh B (2007) Agroforestry systems and practices prevailing in Arunachal Pradesh, their constraints and potentials. In: Puri and Panwar, (eds.). Agroforestry systems and practices. New Delhi: New India Publishing Agency, India.
35. Deb S, Arunachalam A, Arunachalam K, Tangjang S (2007) Traditional homegardens in the humid tropics of North-east India. In: Puri and Panwar, (eds.). Agroforestry systems and practices. New Delhi: New India Publishing Agency, India.
36. Jamini SN, Tikka SBS (2007) Agroforestry practices and systems in North Gujarat. In: Puri and Panwar, (eds.). Agroforestry systems and practices. New Delhi: New India Publishing Agency, India.

37. Harsh LN, Tewari JC (2007) Agroforestry systems in Arid Region (Rajasthan) of India. In: Puri and Panwar, (eds.). Agroforestry systems and practices. New Delhi: New India Publishing Agency, India.
38. Kumar BM, Takeuchi K (2009) Agroforestry in the Western Ghats of peninsular India and the satoyama landscapes of Japan: A comparison of two sustainable land use systems. Sustainability Science 4: 215-232.
39. Sreemannarayana B, Joseph B, Giri Rao LG (2007) Agroforestry systems and practices adopted in Andhra Pradesh. In: Puri and Panwar (eds.). Agroforestry systems and practices. New Delhi: New India Publishing Agency, India.
40. Satish BN, Kushalappa CG (2007) Agroforestry systems of Karnataka: an overview. In: Puri and Panwar, (eds.). Agroforestry systems and practices. New Delhi: New India Publishing Agency.
41. Kumar BM (2007) Agroforestry systems and practices of Kerala. In: Puri and Panwar, (eds.). Agroforestry systems and practices. New Delhi: New India Publishing Agency, India.
42. Paramatma M, Sudhagar RJ, Rathakrishnan P (2007) Agroforestry systems and practices in Tamil Nadu. In: Puri and Panwar, (eds.). Agroforestry systems and practices. New Delhi: New India Publishing Agency, India.
43. Schoeneberger MM (2009) Agroforestry: working trees for sequestering carbon on agricultural lands. Agroforestry System 75: 27-37.
44. Jama B, Kwesiga F, Niang A (2006) Agroforestry innovations for soil fertility management in sub-Saharan Africa: Prospects and challenges ahead. World Agroforestry into the Future: 60.
45. ICAR (2006) Handbook of Agriculture, (5thedn). New Delhi: Indian Council of Agricultural Research.
46. FAO (Food and Agriculture Organization) (2005) Realizing the economic benefits of agroforestry: experiences, lessons and challenges. In: State of the World's Forests 2005.
47. Prasad K (2003) Prospects of Agroforestry in India, XII World Forestry Congress 2003.
48. Sathaye JA, Ravindranath NH (1998) Climate change mitigation in the energy and forestry sectors of developing countries. Annual Review of Energy and Environment 23: 387-437.

49. Jha MN, Gupta MK, Raina AK (2001) Carbon Sequestration: Forest soil and land use management. Annals of Forestry 9: 249-256.

50. Maikhuri RK, Semwa RL, Rao KS, Singh K, Saxena KG (2000) Growth and ecological impacts of traditional agroforestry tree species in Central Himalaya, India. Agroforestry Systems 48: 257-272.

51. Swamy SL, Puri S, Singh AK (2003) Growth, biomass, carbon storage and nutrient distribution in Gmelina arborea Roxb. stands on red lateritic soils in Central India. Bioresource Technology 90: 109-126.

52. Dhyani SK, Newaj R, Sharma AR (2009) Agroforestry: its relation with agronomy, Challenges and opportunities. Indian Journal of Agronomy 54: 249- 266.

53. Chauhan SK, Sharma SC, Beri V, Yadav S, Gupta N (2010) Yield and carbon sequestration potential of wheat (Triticum aestivum) – poplar (Populus deltoides) based agri-silvicultural system Indian. Journal of Agricultural Sciences 80: 129- 35.

54. Sharma R, Sharma E, Purohit AN (1995) Dry matter production and nutrient cycling in agroforestry systems of mandarin grown in association with Albizia anf mixed tree species. Agroforestry systems 29: 165-179.

55. Singh K, Chauhan HS, Rajput DK, Singh DV (1989) Report of a 60 month study on fitter production, changes in soil chemical properties and productivity under Poplar (P.deltoides) and Eucalyptus (E. hybrid) interplanted with aromatic grasses. Agroforestry Systems 9: 37-45.

56. Saha R, Tomar JMS, Ghosh PK (2007) Evaluation and selection of multipurpose tree for improving soil hydro-physical behaviour under hilly eco-system of north east India. Agroforestry Systems 69: 239-247.

57. Gupta N, Kukal SS, Bawa SS, Dhaliwal GS (2009) Soil organic carbon and aggregation under poplar based agroforestry system in relation to tree age and soil type. Agroforestry Systems 76: 27-35.

58. Venkateswaralu J (2010) Rainfed agriculture in India: Research and development scenario. Indian council of Agricultural Research, New Delhi: 508.

59. Samra JS, Singh SC (2000) Silvopasture systems for soil, water and nutrient conservation on degraded land of Shiwalik foothills (sub-tropical northern India). Indian Journal of Soil Conservation 28: 35-42.
60. NRCAF (2007) Perspective Plan Vision 2025 National Research Centre for Agroforestry, Jhansi, Uttar Pradesh.
61. Singh TP (2003) Potential of Farm Forestry in Carbon Sequestration. Indian Forester 129: 839-843.
62. Rai P, Ajit, Chaturvedi PO, Singh R, Singh UP (2009) Biomass production in multipurpose tree species in natural grasslands under semi arid conditions. Journal of Tropical forestry 25: 11-16.
63. Kaur B, Gupta SR, Singh G (2002) Carbon storage and nitrogen cycling in silvipastoral system on sodic soil Northwestern India. Agroforestry Systems 54: 21-29.
64. Narain P (2008) Dryland management in arid ecosystem. Journal of the Indian Society of Soil Science 56: 337-347.
65. CAZRI (2009) Trends in Arid Zone Research in India. In: Amal K, Garg BK, Singh MP, Kathju S, (eds.). Central Arid Zone Research Institute, Jodhpur: 481.
66. Kumar AK (2010) Carbon sequestration: underexploued environmental benefits of Tarai agroforestry. Indian Journal of Soil Conservation 38: 125-131.
67. Hooda N, Gera M, Andrasko K, Sathaye J, Gupta MK, et al. (2007) Community and farm forestry climate mitigation projects: case studies from Uttaranchal, India. Mitigation and Adaptation Strategies for Global Change 12: 1099-1130.
68. Tiwari R, Murthy IK, Killi J, Kandula K, Bhat PR, et al. (2010) Land use dynamics in select village ecosystems of southern India: drivers and implications. Journal of Land Use Science 5.
69. Cairns MA, Meganck RA (1994) Carbon sequestration, biological diversity and sustainable development; Integrated forest management. Environmental Management 18: 13-22.
70. Gladstone WT, Ledig FT (1990) Reducing pressure on natural forests through high yield forestry. Forest Ecology and Management 35: 69-78.

71. Schroeder P, Ladd L (1991) Slowing the increase of atmospheric CO_2: A biological approach. Climate Change 19: 283-290.
72. Sedjo R (1989) Forests: A tool to moderate global warming. Environment 31: 14-20.
73. Shepherd D, Montagnini F (2001) Above-ground carbon sequestration potential in mixed and pure tree plantations in the humid tropics. Journal of Tropical Forest Science 13: 450-459.
74. Ong CK, Swallow BM (2003) Water productivity in forestry and agroforestry. In: W Kijne, R Barker, D Molden (eds.). Water Productivity in Agriculture: Limits and Opportunities for Improvement. CAB International.
75. Murthy IK, Bhat DM, Hegde GT, Nagaraj BC, Murali KS, et al. (2006) Tree Diversity and Carbon Sinks: Role of Village Agro-ecosystems in the Western Ghats, Karnataka, India. International Journal of Agricultural Sustainability, 3: 114-121.
76. MoEF (2010) National Mission for a Green India.
77. Puri S, Nair PKR (2004) Agroforestry research for development in India: 25 years of experiences of a national program. Agroforestry Systems 61: 437-452.
78. Ravindranath NH (2007) Mitigations and adaptation synergy in forest sector. Mitigation and Adaptation Strategies for Global Change 12: 843-853.
79. Garrity D, Okono A, Grayson M, Parrott S (2006) World Agroforestry into the Future Nairobi: World Agroforesty Centre.

Chapter 7

Irrigation Scheduling Research: South African Experiences and Future Prospects

JG Annandale[1], RJ Stirzaker[1, 2], A Singels[1, 3], M van der Laan[1, 3], and MC Laker[1]

[1]Department of Plant Production and Soil Science, University of Pretoria, Private Bag X20, Hatfield, Pretoria 0028, South Africa

[2]CSIRO Land and Water, PO Box 1666, ACT 2601, Australia

[3]South African Sugarcane Research Institute, Private Bag X02, Mount Edgecombe 4300, South Africa

ABSTRACT

Many scheduling approaches have been developed with Water Research Commission funding over the past 4 decades anddeployed with varying levels of success; 2 approaches have won prestigious international awards. Soil-based approacheswhich include measurement of matric potential (tensiometry), water content (neutron probes, capacitance sensors) and depthof wetting (wetting front detectors) have been relatively well accepted by farmers. Atmospheric-based approaches apply,through biophysical modelling of the soil-crop-atmosphere system, thermodynamic limits to the amount of water that canevaporate from a cropped surface under particular environmental conditions. Modelling approaches have been quite empiricalor somewhat more mechanistic, generic or crop specific, with pre-programmed (e.g. irrigation calendars) or real-timeoutput. Novel mechanisms have been developed to deliver recommendations to farmers, including resource-poor irrigators .Although general adoption of objective irrigation scheduling in South Africa is still low, the high cost of electricity andnitrogen, and scarcity of water is reviving the interest of consultants and irrigators in the application of these tools to usewater more efficiently. Where adoption has been relatively high, intensive support and farmer-researcher-consultant interactionshave been key contributing factors. We propose 4 avenues in the R&D domain to ensure responsible water utilisation.Firstly, there is a need to continue to advance existing soil-water measurement technology; and secondly, to further developnew and emerging technologies, like the use of remote sensing. Thirdly, the user-friendliness should be improved as shouldsystems that support existing scheduling tools; and finally, we need to appreciate that farmers are intuitively adaptive managers,and we need to develop simple monitoring tools and conceptual frameworks that enable structured learning.

INTRODUCTION

South Africa receives an average rainfall of 495 mm, well below the global average of 860 mm/a. The total renewable water available per inhabitant is just 1 106 m^3/a, placing South Africa among the driest quintile of countries in the world (FAO, 2005). According to international criteria, over 90% of South Africa is classified as dryland, i.e. arid to sub-humid, with 82% being classified as arid to semi-arid (ARC-ISCW, 2005). Most of the country receives summer rain that is poorly distributed, with droughts being common phenomena (Bennie and Hensley, 2001). Most of the sub-humid to humid areas are non-arable due to steep slopes and/or poor quality soils. Thus, only around 13% (14 x 10^6 ha) of the country is suitable for rain-fed cropping. Irrigation is practised on an estimated 1.5 x 10^6 ha and approximately 0.26 x 10^6 ha are affected by waterlogging and/or salinisation (FAO, 2005). A large proportion of irrigated land is in areas that are too dry for rain-fed cropping, and the area is limited mainly by water scarcity. Irrigated agriculture is by far the biggest user of runoff water in South Africa. In 2000 irrigated agriculture used 7 900 x 10^6 m^3 of runoff, around 61% of the 12 900 x 10^6 m^3 runoff used by all sectors during that year or just under 40% of the estimated 20 x 10^9 m^3 exploitable runoff (DWAF, 2004). Because of this large proportion of South Africa's blue water resources being used by irrigated agriculture, there is understandably a great deal of pressure to transfer water to other sectors. This could have major implications for food security, since a large proportion of several food crops is produced under irrigation, e.g. about 90% of all fruit and vegetables (Nieuwoudt et al., 2004). In view of the scarce water resources and the huge demand by irrigated agriculture, South Africa's Water Research Commission (WRC) has over its 40-year history continuously invested in developing tools that can assist water managers and growers to optimise irrigation water-use efficiency. The primary aim of irrigation scheduling is to minimise wasteful losses of water (percolation beyond what is necessary for salt leaching, surface runoff and evaporation) and maximise transpiration, which is the beneficial loss of water due to its direct link with dry matter production (Tanner and Sinclair, 1983). Scheduling, therefore, plays a fundamental role in determining crop water productivity (CWP), which is a performance indicator used to describe the relationship between water applied and agricultural product output. The pioneering work of, *inter alia*, Gardner (1960), Denmead and Shaw (1962) and Passioura (1988), and set the fundamental theoretical framework of soil-plant-water relations. Much of the research over the past few decades has been to operationalise this framework, i.e. to make it useful to irrigation farmers. This has involved the characterisation of soil-water holding properties, the development of soil-water monitoring tools and methods to predict plant-water use. Although enormous progress has been made, and the scientific elements of the irrigation scheduling package appear to be in place, adoption by irrigators remains limited in South Africa and internationally (Leib et al., 2002; Stevens et al., 2005; Stirzaker, 2006). A survey of 332 irrigation schemes in South Africa by Stevens et al. (2005) indicated that objective scheduling was being applied by 18% of farmers only, with the rest relying on approaches based on 'instinct, knowledge, experience and confidence gained over many years of farming'. In practice, irrigators often irrigated with fixed amounts or at a constant interval with little regard to variability in weather conditions and actual crop-water requirements. The objective of this review is to document irrigation scheduling-related research over the past 4 decades in South Africa, with the main emphasis on the advances made through WRC-funded irrigation scheduling research and technologytransfer projects. The aim is to highlight technical achievements of soil- and atmospheric-based scheduling approaches, to reflect on why adoption of new technology has been much lower than expected, and to consider possible approaches needed to take irrigation scheduling forward in the future.

AN OVERVIEW OF WRC-FUNDED IRRIGATION

Scheduling and Related Research in South Africa

During the late 1970s and early 80s, WRC-funded research focused on allowable depletion, or the so-called 'profile available water capacity' (PAWC) concept, and models to derive PAWC values for different crop-soil combinations. There was also research during this early period on irrigation scheduling of agronomic, vegetable and pasture crops. By the late 1980s, the focus had shifted to atmospheric-based soil-water balance modelling, with much of the results of this work put into practice by the late 1990s, with the development of a variety of computer-based scheduling approaches. The turn of the millennium saw the first of the social science research looking at technology adoption issues. Thereafter there was continued development of existing methods and their deployment in a variety of specific applications with greater focus on technology transfer. Lastly there was an attempt to develop irrigationscheduling tools for the small-scale irrigation sector. Table 1 shows a list of final reports emanating from WRCsponsored research projects on irrigation scheduling published since the early 1980s, chronologically representing efforts and focus areas over the years. Most of these reports can be accessed and downloaded free of charge from the WRC's website (www.wrc.org.za). While a comprehensive review of this vast body of research is beyond the scope of this article, selected research that has enhanced our knowledge and the approach of researchers to assist irrigators in devising more effective scheduling programmes is discussed below.

Table 1: Details of final reports of irrigation scheduling related research projects sponsored by the Water Research Commission

Year	Report Title	Report No.	Reference
1982	The Determination of the Profile Available Water Capacities of Soils	Univ. Fort Hare Report	Hensley and De Jager (1982)
1983	Water Use Efficiency of Irrigated Crops as Influenced by Varying Cultivation Practices and Root Configurations	UOFS Report	Botha et al. (1983)
1985	The Development of Profile Available Water Capacity Models	98/1/85	Boedt and Laker (1985)
1987	Research on a Weather Service for Scheduling the Irrigation of Winter Wheat in the OFS	117/1/87	De Jager et al. (1987)
1987	Evapotranspiration and Water Use Studies in Wheat and Soybeans with the Help of the Weighing Lysimeter Technique	112/1/87	Meyer et al. (1987)
1988	Water Balance Model for Irrigation Based on Soil Profile Water Supply Rate	144/1/88	Bennie et al. (1988)
1989	Research on Improving Irrigation Management based on Soil Water Monitoring and Detailed Knowledge of Profile Available Water Capacities	166/1/89	Vanassche and Laker (1989)

1989	Correction factors for evaporimeter coefficients used for scheduling irrigation of wheat	151/1/89	Van Zyl et al. (1989)
1990	Drupbesproeiing by Tamaties (Drip Irrigation of Tomatoes)	185/1/90	Fischer and Nel (1993)
1993	Practical Scheduling of Irrigation in the Northern Transvaal	152/1/93	Burgers and Kirk (1993)
1994	Research on the Climatic Dependence of Evaporation Coefficients	260/1/94	Van Zyl and De Jager (1994)
1994	Research on Maximising Irrigation Project Efficiency in Different Soil- Climate-Irrigation Situations	226/1/94	Mottram and De Jager (1994)
1994	Waterverbruik en Waterverbruiksdoeltreffendheid van Gematigde Aanplante Weidings onder Bespoeiing (Water Use and Water Use Efficiency of Temperate Planted Pastures under Irrigation)	257/1/94	Steynberg et al. (1994)
1995	Proceedings of the Southern African Irrigation Symposium, Durban, 1991	TT 71/95	Water Research Commission (1995)
1996	The Response of Citrus Seedlings to Soil Compaction And Variations In Soil Water Potential In the Upper Range of Plant-Available-Water	261/1/96	Mkhize et al. (1996)
1996	Die Fasilitering van Tegnologie Oordrag deur Verbeterde Besproeiings- riglyne vir Groente en 'n Meganistiese Gewasmodelleringbenadering (Facilitating Technology Transfer through Improved Irrigation Guidelines of Vegetables and a Mechanistic Crop Modelling Approach)	476/1/96	Annandale et al. (1996)

1997	Reaksie van Gewasse op Voorafgeprogrammeerde Tekortbesproeiing (Reaction of Crops to Pre-Programmed Deficit Irrigation)	423/1/97	Bennie et al. (1997)
1998	Use of Computer Models for Agricultural Water Management on Ecotope Level	102/98	Bennie et al. (1998)
1999	SAPWAT: A Computer Program for Establishing Irrigation Requirements and Scheduling Strategies in South Africa	624/1/99	Crosby and Crosby (1999)
1999	Facilitating Irrigation Scheduling by Means of the Soil Water Balance model	753/1/99	Annandale et al. (1999a)
2000	Factors Which Influence the Acceptability of Irrigation Scheduling with Specific Reference to Scheduling Models	893/1/00	Botha et al. (2000)
2000	An investigation of the Stem Steady State Heat Energy Balance Technique in Determining Water Use by Trees	348/1/00	Savage et al. (2000)
2000	Irrigation Requirements of Selected Crops under Small-Scale Production: Linking On-Farm and On-Station Research	689/1/00	Walker (2000)
2001	Research on a Computerised Weather Based Irrigation Water Management System	581/1/01	De Jager et al. (2001)
2002	The Development of an Integrated Information System for Irrigation Water Management Using the WAS, SWB and RISKMAN Computer Models	946/4/02	Benade et al. (2002)
2002	Water Use and Water Use Efficiency of Fodder Crops under Irrigation: Part 1— Annual Subtropical Crops	573/1/02	Marais et al. (2002)

2002	Two Dimensional Energy Interception and Water Balance Model for Hedgerow Tree Crops	945/1/02	Annandale et al. (2002a)
2003	Optimization of Irrigation Management in Mango Trees by Determination of Water and Carbon Demands to Improve Water Use Efficiency and Fruit Quality	1136/1/03	Pavel et al. (2003)
2003	The Selection and Calibration of a Model for Irrigation Scheduling of Deciduous Fruit Orchards	892/1/03	Volschenk et al. (2003)
2003	Deficit Irrigation Studies to Improve Irrigation Scheduling in Deciduous Fruit Orchards	892/2/03	Beukes et al. (2003)
2004	Building Capacity in Irrigation Management with Wetting Front Detectors	TT 230/04	Stirzaker et al. (2004)
2005	Technology Transfer of the Soil Water Balance (SWB) Model as a User- Friendly Irrigation Scheduling Tool	TT 251/05	Annandale et al. (2005)
2005	The Range, Distribution and Implementation of Irrigation Scheduling Models and Methods in South Africa	1137/1/05	Stevens et al. (2005)
2007	Predicting the Environmental Impact and Sustainability of Irrigation with Coal Mine Water	1149/1/07	Annandale et al. (2007)
2008	Real Time Irrigation Advice for Small-Scale Sugarcane Production Using a Crop Model	1576/1/08	Singels and Smith (2008)
2009	Increasing Water Use Efficiency of Irrigated Sugarcane by Means of Specific Agronomic Practices	1577/1/09	Olivier et al. (2009)

2010	Standards and Guidelines for Improved Efficiency of Irrigation Water Use from Dam Wall Release to Root Zone Application: Guidelines	TT 466/10	Reinders et al. (2010)
2010	Wetting Front Detector Transfer of Technology	K8/599/4	Stevens and Stirzaker
2010	Adapting the Wetting Front Detector to Small-Scale Furrow Irrigation and Providing a Basis for the Interpretation of Salt and Nutrient Measurements from the Water Sample	1574/1/10	Stirzaker et al. (2010b)

The Concept of Plant-Available Water

Knowing when to irrigate, i.e. the optimum stage in the drying cycle at which to apply water, and how much plant-available water the soil profile can hold, can assist an irrigator in improving irrigation water-use efficiency. For the determination of allowable depletion or 'PAWC', as it was defined by South African researchers, the correct determination of both the upper and lower limits is required (Bennie, 1995; Hensley and De Jager, 1982.) Since the soil-water potential at 'field capacity' differs widely between soils, a single soil-water potential value cannot be used for all soils (Bennie, 1995; Hensley and De Jager, 1982; Ratliff et al., 1983). For the fine sandy soils at Vaalharts (about 8% to 10% clay and a sand fraction highly dominated by fine sand), 'field capacity' is, for example, at a soil-water potential of about -5 kPa and much of the plant-available water is held at high soil-water potentials, causing these soils to have unexpectedly high PAWC values of more than 100 mm·m-1 for crops like maize and wheat, Thus, as in the USA and elsewhere, it was realised that *in situ* determined 'drained upper limits' (DUL) had to be used. Boedt and Laker (1985) and Bennie et al. (1988) carried out several DUL determinations from which they derived remarkably similar regression equations for estimating DUL from the silt plus clay contents of different depth intervals of a soil profile (Bennie, 1995). Hensley and De Jager (1982) defined the lower limit of PAWC using plant parameters; a point Hensley termed 'first material stress' (FMS). This is not to be confused with the more commonly used international lower limit for plantavailable water (PAW) of permanent wilting point, a much lower soil-water content, typically at a matric potential of around -1.5 MPa. Based on earlier work in KwaZulu-Natal, Hensley used wilting of maize between 10:00 and 10:30 as a visual indicator of stress (Hensley and De Jager, 1982). This worked well under the climatic conditions prevailing at Fort Hare, but during periods of extreme evaporative demand (Class A pan readings of up to 14 $mm \cdot d^{-1}$) at Vaalharts, it could not be used. Under such conditions, maize wilted at 10:00 even on soils that were at field capacity (Boedt and Laker, 1985). Pre-dawn leaf-water potential, measured by means of a pressure chamber, was found to be a good indicator of FMS under normal conditions. Under conditions of extreme evaporative demand, however, its results also became very erratic (Boedt and Laker, 1985; Laker et al., 1987). Using the results of Hensley and De Jager (1982), Laker (1982) developed a model by means of which PAWC could be estimated accurately for different soils, using very simple soil properties. Boedt and Laker (1985) followed this up with PAWC studies with maize, wheat, cotton and peas on an even wider range of soils in 3 different regions, i.e. at Fort Hare (Eastern Cape), Vaalharts (Northern Cape) and Loskop (Mpumalanga). All PAWC estimates were for plants at full canopy and full root-system development. Vanassche and Laker (1989) determined PAWC at various growth stages for wheat, durum wheat and maize. Bennie and his co-workers made use of these concepts in the development of the BEWAB irrigation model (Bennie, 1995; Bennie et al., 1988; Bennie et al., 1997). A significant finding in the various PAWC studies was that field crops could extract water to great depths and that PAWC was larger than had been assumed previously, especially in the fine sandy soils of Vaalharts (Boedt and Laker, 1985). Consequently intervals between irrigations could be stretched and evaporation losses were less than with more frequent irrigations. Flood irrigation was used in the experiments, but Bennie and his co-workers developed strategies for using it under overhead systems as well (Bennie, 1995). Scheduling irrigation in this way increased irrigation water-use efficiency, without compromising crop yields, compared with controls that were irrigated at shorter intervals, maintaining higher levels of plant-available water in soils (Hensley and De Jager, 1982; Boedt and Laker, 1985; Vanassche and Laker, 1989). This was in line with the findings of similar studies in the USA at the same time (e.g. Buchheim and Ploss, 1977). An additional innovation studied during the PAWC research by Vanassche and Laker (1989) was 'deficit highfrequency irrigation', an irrigation-scheduling approach that was pioneered in the USA (e.g. Miller, 1977; English and Nakamura, 1982). In this approach the water content of the soil is not allowed to drop below the point of first material stress, i.e. the crop is never stressed. With 'deficit' is meant that the entire soil profile soil is not filled to DUL during an irrigation event – after initially starting on a full profile. Reasons for improved irrigation water-use efficiencies were extraction of more water from the deep subsoil; and leaving room for effective use of rain water if a rainfall event occurred shortly (within a few days) after irrigation (this is not relevant to arid areas). Vanassche and Laker (1989) studied different variations of this type of deficit irrigation with maize and wheat, e.g. irrigation is applied each time that 75% of PAWC has been extracted, i.e. leaving 25% of PAWC, but enough water is applied to replenish it up to only 75% of PAWC and not up to full DUL. This approach generally gave better irrigation water-use efficiencies than other approaches, as was also found in the USA. Laker (1985) discussed how knowledge of PAWC could be used to reduce design peaks so that various irrigation systems and systems designed with smaller capacities (thus being less expensive) could cope with crop-water requirements during peak demand periods. He also indicated how this could be implemented under, for example, centre-pivot irrigation, where a soil profile cannot be filled with a single large irrigation event – as under flood irrigation. Starting with a dry profile, the soilwater content could be built up by 'over-irrigating' in the early part of the season to have a full profile at the start of the peak demand period. By means of 'deficit high frequency' irrigation the soil water would then be extracted gradually during the peak demand, to end with a profile at FMS at the end of the peak period. Bennie and his co-workers devised different variations of this suggested approach and had great success with it in practical irrigation scheduling of field crops (Bennie, 1995; Bennie et al., 1997). It was also applied successfully to some horticultural crops, such as tomatoes (Fischer and Nel, 1990; Fischer, 1995), and at farm scale with grapevines (Nel, 1995b). The latter was introduced during a period of severe water restrictions. A computerised irrigation system was used. A 'start irrigation' level was set at a soil-water content halfway between the upper and lower limits of PAWC. A 'stop irrigation' level was at a deficit

30 mm below the upper limit. This left a reserve capacity of 30 mm for interception and effective storage of rainfall, 'a most important consideration when the effective use of available irrigation water is to be maximised' (Nel, 1995b). It is important to keep in mind that in the case of leafy vegetables deficit irrigation usually leads to significant yield losses, as for example found by Van Averbeke and Netshithuthuni (2010) with Chinese cabbage in a WRC-sponsored study. The shallow root system and the fact that a high water supply is required for maximum leaf growth means that very frequent full irrigations are required for these crops, i.e. the water content of the soil must be kept close to DUL the whole time. During the past decade or more 'regulated deficit irrigation' (RDI) and 'partial root zone drying' (PRD) approaches have been widely researched as deficit irrigation strategies aimed at increasing water-use efficiencies in areas with water scarcity (reviewed by e.g. Costa et al., 2007). These have also been implemented successfully in practice. RDI is a method of stressing a crop at a specific growth stage (or stages) to control excessive vegetative growth without affecting fruit yield (Grant, 2000; Costa et al., 2007), so as to improve water-use efficiency and/or crop quality. In PRD irrigation is applied to half the root zone, while the other half is allowed to dry (Grant, 2000; Costa et al., 2007). It is implemented by irrigating alternative inter-rows (e.g. in vineyards), whereupon the dry side is irrigated, and the previously irrigated side is allowed to dry. RDI and PRD are particularly effective practices for grapevines and various deciduous fruit crops (Grant, 2000; Costa et al., 2007). Water-use efficiency (WUE) increases of up to over 60% or 70% without harming crop yield or quality have been recorded in best-case scenarios (Costa et al., 2007). In South Africa the use of RDI in deciduous fruit crop irrigation has, *inter alia*, been studied by Beukes et al. (2003). PRD has recently been studied intensively at the ARC-Infruitec Nietvoorbij as potential strategy for irrigating wine grapes in South Africa. This has been reported in several publications, including those by Lategan and Howell (2010) and Myburgh (2011a). RDI and PRD give variable results with evergreen fruit crops (Costa et al., 2007). Mangoes are interesting in the sense that deficit irrigation not only achieved higher WUE (Pavel and De Villiers, 2004), but also produced larger and better quality fruit than full irrigation (Spreer et al., 2007). Additional studies with deficit irrigation strategies have been conducted on various other types of crops (Costa et al., 2007). Potatoes, for example, generally gave

negative results, because of the shallow root system, but in a best-case scenario irrigation water use was increased by 60% without lowering yield. Various types of vegetables generally gave negative responses to deficit irrigation. There appears to be great scope and an urgent need for more research on deficit irrigation strategies in South Africa, considering different approaches and studying a variety of important crops.

MEASURING TECHNIQUES FOR IRRIGATION SCHEDULING

In order to implement different irrigation-scheduling strategies efficiently, measurements need to be made. Types of measuring techniques can be broadly classified as plant, soil or atmospheric based.

Plant-Based Techniques

Several plant-based measuring techniques have been studied for their potential use in irrigation scheduling. The 'pistol' type infrared thermometer for measuring canopy temperature received much attention as a potential practical tool for irrigation scheduling, because of its apparent ease of use (Reginato, 1995), but determining water stress from canopy and air temperatures proved problematic because of the effects of extreme atmospheric conditions, e.g. situations of very low vapour saturation deficit (VSD). A rather more complex 'crop water stress index' (CWSI) taking VSD into account had to be developed to complement this device (Idso et al., 1981; Jackson et al., 1981). Measurement of leaf-water potential has received by far the most attention of the plant-based measurement techniques in irrigation scheduling. Savage and his co-workers at the University of KwaZulu-Natal are internationally renowned for their studies on thermocouple psychrometers as method to determine leaf-water potential (e.g. Savage and Wiebe, 1987). However, Scholander-type pressure chambers have been most widely used for determining leaf-water potential, including in South Africa (e.g. Hensley and De Jager, 1982; Boedt and Laker 1985; Vanassche and Laker, 1989; Nel, 1995a). Although midday determination of leaf-water potential has been used successfully (e.g. Nel, 1995a), pre-dawn leaf-water potential under normal conditions gives the best indication of the onset of water stress in plants (Hensley and De Jager, 1982; Boedt and Laker, 1985, Vanassche and Laker, 1989; Laker, 2004) because shortly before dawn, leaf-water potential equilibrates with soil-water potential (Laker, 2004). The exception is under extremely hot, dry conditions, such as often experienced in summer in the central irrigation areas of South Africa, conditions under which pre-dawn leaf-water potential measurements were shown to be unreliable (Boedt and Laker, 1985; Laker et al., 1987; Vanassche and Laker, 1989). Similar problems have been found with other plant-based measurements also under extreme conditions (Reginato, 1995). Research using pre-dawn leaf-water potential by means of Scholander-type pressure chambers has produced threshold values that have facilitated its use in practical irrigation scheduling on a limited scale, for example in the wine-grape industry in the Western Cape. Some irrigation scheduling consultants provide advice using pressure chambers, and a few wine estates also operate their own instruments (Myburgh, 2011b). Many other plant-based measurements have been researched in South Africa (e.g. stomatal conductance, CO_2 exchange rate, sap flow, leaf extension rate), but none of these have evolved into practical irrigation-scheduling methods.

Soil-Based Techniques

Measurement of soil-water status has a long history, from the tensiometers developed in the 1930s (Richards and Neal, 1936), to the neutron probe in the 1950s (Gardner and Kirkham, 1952), through to a whole range of tools based on the measurement of the dielectric properties of water in soil (Charlesworth, 2005). Perhaps the most familiar tool to irrigators is the tensiometer, which measures matric potential (suction) at the wet end of the plant-available water spectrum, making it valuable for horticultural crops. Streutker (1978) used tensiometers to keep the top (0 cm to 45 cm) soil layer at a high water potential and the subsoil water at a low water potential, achieved through medium frequency irrigation. This early research demonstrated how objective scheduling could result in significant water savings with no loss of yield. Streutker et al. (1981) promoted the use of tensiometers in the Loskop Irrigation District during the 1980s, and was very successful at improving irrigation management by, amongst other interventions, highlighting the contribution of shallow water tables to crop-water use. They developed water-yield functions for several crops enabling farmers to benchmark their water productivities and gauge whether or not they had room to improve. The neutron probe has long been the standard instrument for measuring soil-water content. It has been used with great success in practical irrigation management, especially in orchards and vineyards (Mkhize et al., 1996; Nel, 1995b). On a large commercial citrus estate near Marble Hall, Mkhize et al. (1996) were able to demonstrate 24% water saving, while increasing yield and fruit quality through neutron-probe scheduling following a PAWC approach, thereby facilitating a significant saving in irrigation costs. The neutron probe is, however, being superseded by a range of logged capacitancetype sensors, such as DFM capacitance probes (Haarhoff, 2011). Although potentially not as representative as neutron probes because of their smaller volume of measurement, automated capacitance sensors are usually far easier to deploy in farmers' fields. The wetting front detector (WFD) was conceived and developed against the background of poor adoption of commonly available technologies. Essentially the WFD reframed the age-old irrigation scheduling question from 'when to turn http://dx.doi.org/10.4314/wsa.v37i5.12 Available http://www.wrc.org.za ISSN 0378-4738 (Print) = Water SA Vol. 37 No. 5 WRC 40-Year Celebration Special Edition 2011 756 ISSN 1816-7950 (On-line) = Water SA Vol. 37 No. 5 WRC 40-Year Celebration Special Edition 2011 the water on' to 'when to turn it off' (Stirzaker, 2003). The focus of soil-based monitoring had been on specifying refill points, i.e. how dry the soil could be allowed to get without affecting production of the crop. Long intervals between irrigation events meant that sprinklers could be moved less often; however, the advent of micro-irrigation changed this perspective. Irrigation could occur at any time, and was often carried out daily or pulsed several times per day. WFDs are based around the tipping-bucket analogy, where soil layers are viewed as a sequence of buckets that store water. As the upper bucket is filled by irrigation, it tips and spills excess water into the bucket below and so on down the profile. The WFD was designed to show when water moved from one layer to the next. It is comprised of a specially shaped funnel, a filter, and a float plus indicator mechanism. The funnel shape was designed so that the soil at its base reaches saturation when matric potential of the soil outside the funnel is around 2 kPa to 3 kPa (Stirzaker

2008), which corresponds to a relatively 'strong' wetting front. Once saturation occurs at the base of the funnel, free water flows through a filter into a small reservoir and activates a float. The float trips a magnetically latched indicator, visible to the irrigator. These detectors are often installed at different depths and used in a similar way to that for tensiometry, namely a shallow detector indicating water entering the root zone and a deeper detector possibly warning of over-irrigation (Stirzaker and Hutchinson 2005). Fessehazion et al. (2011) used WFDs and the principles of adaptive management to manage excess nitrogen use in irrigated pastures. In this case, the depth of wetting and nitrate concentrations recorded in WFDs were measurements that broadly integrated many of the processes involved in the soilwater balance and N cycle. Even simple thresholds for action were able to reduce N application by up to one third, whilst retaining yield levels and improving the quality of the pasture in which this application was tested.

Atmospheric-Based Approaches

Atmospheric-based approaches apply, through biophysical modelling of the soil-crop-atmosphere system, thermodynamic limits to the amount of water that can evaporate from a cropped surface under a particular set of environmental conditions. Early work was based on evaporation measured from a pan together with crop coefficients which were then used to estimate evapotranspiration (*ET*) (Doorenbos and Pruitt, 1977). The standard Class A-evaporation pan was most widely used in South Africa, and was reported by Stevens et al. (2005) still to be popular amongst commercial farmers and consultants in the Breede Water Management Area. Customised approaches, such as the Scheepers and Vaalharts pan, have also been successfully used in the past, but only a limited number of Northern Cape farmers still use the Vaalharts pan to measure evaporation (Stevens et al., 2005). The aptly named 'Green Book' (Green, 1985) made extensive use of crop factors and evaporation pan data to estimate crop-water requirements of several crops throughout South Africa and was widely used, particularly for planning purposes. The evaporation pan/crop coefficient approach was later shown to have serious limitations (Van Zyl et al., 1989). Panand crop-evaporation coefficients depend strongly on climate, and the use of constant values can lead to inaccurate estimates of crop evaporation (De Jager and Van Zyl, 1989; Van Zyl and De Jager, 1992; Annandale and Stockle, 1994). De Jager and Van Zyl (1989) proposed a breakdown of the single crop coefficient into vegetation and soil sub-components. Each of these was further broken down to account for canopy cover and plant- and soil-water status. De Jager and Van Zyl (1989) used evaporation from a reference crop to represent atmospheric evaporative demand. This theoretical formulation of crop evaporation provided the basis for weather-based water-balance models used in many irrigation- scheduling tools and crop models. Van Zyl and De Jager (1987) and Van Zyl et al. (1990) demonstrated the accuracy of the Penman-Monteith equation (the energy balance of a cropped surface), for estimating evaporation from an unstressed wheat crop and determined typical values of canopy conductance, a key plant parameter in this equation. Although the Penman-Monteith equation requires detailed weather data as input, Van Zyl and De Jager (1987) demonstrated that a modified method using sunshine duration, temperature and a Piché evaporimeter produced reliable evaporation estimates as well. This, and the development of automated weather stations, paved the way for the Penman- Monteith method to replace the evaporation pan/crop coefficient approach as a standard method to estimate crop-water use in South Africa at the time. As automated weather stations became more common, the high-frequency measurement (every few seconds) of air temperature, vapour pressure deficit, wind speed and solar radiation, made the calculation of the evaporation from a reference crop surface based on energy balance principles, commonplace. The international community has settled on a standardised version of the Penman-Monteith equation, that assumes a full cover, well-watered, 12 cm tall reference crop, with an albedo of 0.23 and canopy resistance of 70 s·m-1 (Allen et al., 1998), and this has also been adopted in South Africa.

Modelling the Soil-Water Balance

All approaches to modelling the soil-water balance have empirical and mechanistic aspects to them. Models are either crop-specific or can be described as 'generic' in nature if they can be used for several crops, with pre-programmed (e.g. irrigation calendars) or real-time output. Four WRC-supported irrigation-scheduling modelling efforts stand out. These are BEWAB, PUTU, SWB and My Canesim. Bennie et al. (1988) determined long-term average irrigation water requirements and efficient water-management strategies for various crops (wheat, maize, groundnuts, cotton and peas) for irrigation schemes in the semi-arid regions of South Africa (Vaalharts, Sander, and Ramah). This new knowledge was incorporated into a software package named BEWAB (BEsproeiingsWAterBestuursprogram – Afrikaans for Irrigation Water Management Program) that was widely adopted by farmers. The package provided seasonal irrigation requirement and pre-plant schedules of irrigations for specific crop/site/soil/planting date scenarios. It also related irrigation requirements and schedules to a user-specified target yield through linear production functions derived from experimental data. Strydom (1998) incorporated transpiration efficiency concepts from Tanner and Sinclair (1983) into BEWAB, thus making it applicable to any site for which maximum biomass yield and maximum evapotranspiration information were available. The package suggested appropriate values for these inputs for different locations around South Africa. The concepts developed by De Jager et al. (1987) were incorporated into the PUTU wheat model that was used to provide weekly irrigation scheduling advice to wheat farmers at Vaalharts. Electronically recorded weather data were downloaded via telephone modems and landlines. This was particularly impressive considering that these were manual telephone exchanges. The model was run by the research team to generate crop- and soil-water status information (e.g. expected onset of water stress, water use of the past 7 days, deep drainage out of the root zone, expected date of next irrigation and current soil-water status). This information was faxed to the local extension officer, who conveyed it to farmers when they phoned. This approach to advice dissemination was not very successful. The project highlighted the need for personal interaction with farmers and for marketing of the service to ensure better uptake. The experience gained from this research also led to the development of a generic crop model (PUTU-any crop) that was used to determine water and irrigation requirements of various crops (e.g. De Jager and Singels, 1994; De Jager et al., 2001). It was used by irrigation consultants and extension officers for weather-based irrigation scheduling at Bergville and Winter ton in KwaZulu-Natal and on the Rite River and other Free State/Northern Cape irrigation schemes, as well as in Limpopo Province at ZZ2 on tomatoes and other crops (De Jager and Mottram, 1996; De Jager and Kennedy, 1996). The SWB model is a mechanistic, generic crop model which was developed as a real-time irrigation-scheduling tool (Annandale et al., 1999a). Evapotranspiration is calculated according to the Penman-Monteith grass reference method as recommended by the Food and Agricultural Organization (FAO) (Allen et al., 1998). The soil-water balance can be modelled using either a cascading soil-water balance or a finite difference model (Annandale et al., 1999a). Daily crop dry matter accumulation was taken as the lower value of either radiation-limited growth (Monteith, 1977) or water-limited growth (Tanner and Sinclair, 1983). Thermal time was used to calculate phenology and partitioning with the effect of water stress accounted for through the use of a stress factor. The crop- and soil-water components of the model have undergone extensive parameterisation and testing for commonly grown South African crops including potatoes, sunflower, maize, soybean, canola, fescue, lucerne and a wide range of vegetable crops (Jovanovic and Annandale 1999; Jovanovic et al., 1999; Annandale et al., 2000; Jovanovic and Annandale 2000a; Jovanovic et al., 2000b; Jovanovic et al., 2002; Tesfamariam, 2004; Annandale et al., 2007). An FAOtype crop factor approach was also included in the model to include crops for which limited data were available and for tree crops which cannot be 'grown' using the simple crop model (Jovanovic and Annandale, 1999). Singels and Smith (2009) developed and implemented a weather-based irrigation scheduling advice service for sugarcane growers. The MyCanesim system (Fig. 1, described by Singels, 2007) was automatically executed daily on behalf of subscribers and advice was made available through web downloads for extension staff and commercial farmers, or by cellular text messages for small-scale farmers. More details on the 4 irrigation-scheduling models described briefly above can be found in the references given.

ADOPTION

The research projects described above produced many novel mechanisms that were successfully applied by the target endusers. Yet when the irrigation industry is viewed as a whole, the picture is not reassuring. Stevens et al. (2005) showed that

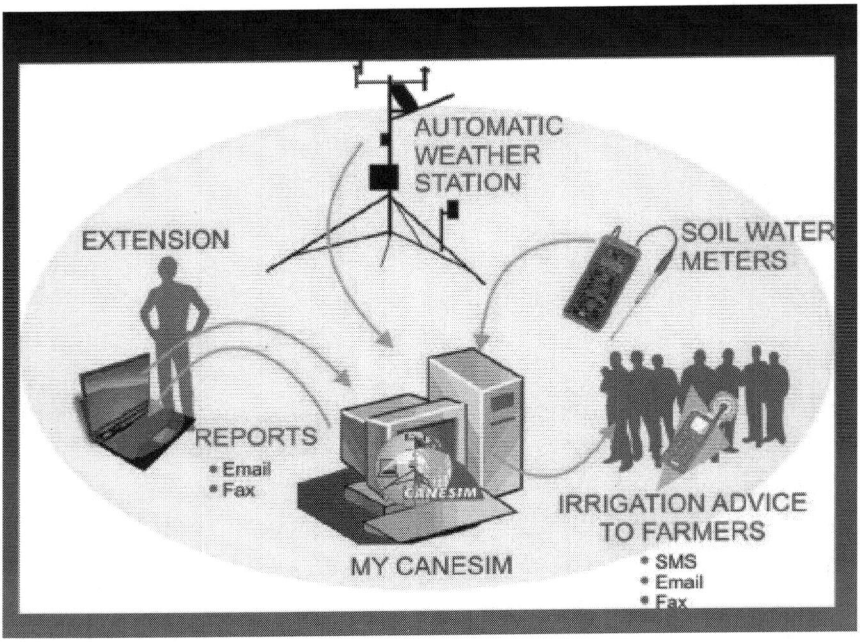

Figure 1 *Schematic presentation of MyCanesim irrigation-schedulingsystem*

Only 18% of commercial irrigators used the products of science to help them schedule irrigation. Their national survey covered 332 irrigation schemes and included semi-structured interviews with irrigation professionals (consultants, advisors, industry experts and irrigation specialists), followed by a more quantitative survey of large- and small-scale farmers to better understand human factors influencing the adoption of objective scheduling. Findings included that scheduling method was closely related to irrigation system, with users of mechanized systems (pivot, micro, drip) showing a positive view towards objective scheduling, and users of flood irrigation showing little interest. The technology level, size of the farm and value of crops grown were also found to influence adoption of more sophisticated scheduling methods. Flat-rate water tariffs on the majority of schemes provided little incentive to schedule objectively, while unreliable water delivery also hindered adoption. In many cases water was one of the cheapest inputs into the irrigation business, and so applying luxury quantities was considered a cheap form of insurance (Stirzaker, 1999). This could be an acceptable strategy in situations where over-irrigation is not considered problematic. However, in many situations injudicious over-irrigation can have serious negative impacts, such as, for example, raised water tables (often leading also to salinisation), high drainage costs, nutrient leaching, crop yield reductions and/or soil or human-health impacts. In response to low adoption of the SWB model, the WRC commissioned a Technology Transfer project to train potential users at the national scale and to make software changes in order to improve user-friendliness. Annandale et al. (2005) reported that these training exercises were largely successful in general knowledge capacity building, but did not improve the adoption of the model itself. Many of the consultants were not irrigation advisors *per se*, but used the course to update and refresh their knowledge on plant/water relations. Others continued to use their own models or tools, and this was attributed to the fact that they were more familiar with their own systems and did not see enough benefit in changing over to another system. Some felt that the model was still too complex and required too many input parameters. BEWAB, on the other hand, was used by around 500 growers. Its successful adoption was ascribed largely to the credibility of the developer, Alan Bennie, who is a well-respected academic, but with the uncanny ability to translate basic scientific findings into practical solutions for on-farm management. He grew up on an irrigation farm at Vaal harts and is also a successful farmer in the area where it was rolled out. It is probable that the intimate knowledge of the local conditions, including knowledge of the views and attitudes of the local farmers, was instrumental in developing a product that satisfied the particular needs of those farmers in many ways, PUTU was a technology developed ahead of its time. Considering rapid development in communication and computer infrastructure, it might have been more successful had it come 10 years later. The development of My Canesim was focused on the difficulties of getting buy-in from farmers. It was first implemented in 2004 on a small-scale grower scheme in Pongola (Singels, and Smith, 2006), and in 2007, advice via text message was provided to 45 small-scale growers. Most of the farmers indicated that they found the advice useful and that it helped them to better understand the value of scheduling. Frequent face-to-face interaction with farmers was observed to be essential in ensuring that advice was understood and implemented. Reliable feedback on irrigation actions and rainfall was also needed to ensure accurate simulation and relevant advice. The approach followed was to provide direct, simple advice and not to confront users with the complexity of the system. This work received recognition by being awarded the International Commission on Irrigation and Drainage (ICID) WATSAVE Award for Innovative Water Management in 2007. Additional work is envisaged to further improve as well as implement this system more widely in the South African sugar industry. The wetting front detector (WFD) was essentially pioneered in South Africa and the research and development effort was framed by the ongoing problem of poor adoption. After a minimum of 12 months' evaluation of prototype detectors, 54 irrigators, or their advisors who participated in the study, were surveyed. All participants perceived the device as simple and intuitive, and based on their own experience, 82% believed it conferred a relative advantage over what they had been doing (Stirzaker et al., 2010a). Those experiencing problems were either using furrow or centre-pivot irrigation, methods not ideally suited to the WFD. In 2003, the team that developed the WFD were awarded the ICID WATSAVE award for 'outstanding contribution to water saving and water conservation in agriculture'. The following year the commercial version was released by a South African irrigation company. Over 15 000 detectors were sold over the next 7 years, mostly in South Africa and Australia but also in Argentina, Chile, Peru, Spain and Greece. Yet the momentum of the early success has been difficult to maintain. Although a simple device, it turned out that a good deal of local knowledge was required to get the best out of the detectors. In particular, the depth of placement was crucial and the need to fit the soil type and irrigation method. Despite the sales volumes, the business model did not generate the returns that allowed advisors or consultants to learn and adapt and solve new problems as they arose. Successful adoption is far more than just the technology. The distribution network, back-up assistance and availability of spare parts are just as important. Adoption of irrigation scheduling among the South African small-scale irrigation schemes has been particularly poor (Fanadzo et al., 2010). Water-limited crop productivity in small-scale irrigation schemes has been attributed to socio-economic, political, climatic, edaphic and design factors (Bembridge, 2000), and has been exacerbated by dilapidated irrigation equipment and poor farmer performance (e.g. poor weed and fertiliser management) (Crosby et al., 2000; Fanadzo et al., 2010). In such cases, it would be important to simultaneously address factors limiting primary production, whilst committing resources to irrigation-scheduling initiatives. Poor adoption, therefore, remains a conundrum for those whose mandate it is to improve water-use efficiency. Eighteen years ago, Burgers and Kirk (1993) wrote '...computers are fast becoming an essential facility on farms, and a user-friendly program which computes the amount of water to be applied on each of the scheduled dates, may well be the answer to the farmer's scheduling problems'. This prediction has not materialised, and the use of computer models for irrigation scheduling, especially by farmers themselves, has remained low (Botha et al., 2000). Similarly, it was expected that once soil-water monitoring equipment was more affordable and user-friendly, widespread adoption would follow naturally. Yet adoption seems to require a dedicated service as well. For example, the Orange-Vaal Water Users Association, managed by Griqualand Wes Kooperasie (GWK), includes around 91 000 ha of irrigated land (about 2 600 centre-pivots) of which more than two-thirds is objectively scheduled, 55% to 65 % by the GWK service, which makes use of 2 systems (Haarhoff, 2011). The older of the two is a simple neutron probe system on about 200 pivots, where farmers are sent weekly updates of soil-profile-water content via e-mail, fax or text message. The successful implementation of this system over many years is attributed to it being designed and customised around meeting the specific needs of farmers for the region (Crosby, 2004). The new technology used by GWK on about 1 300 pivots uses DFM capacitance probes and Irricheck software that enables the farmer to make his own scheduling decisions. However, a service is still provided in that GWK offers support through field observations for probe calibration (Haarhoff, 2011). Lower than expected adoption does not, in itself, question the value of the research effort. For example, SWB was originally conceived as a tool for farmers, but had more positive spin-offs in other applications. The model has been effectively used as a teaching tool in undergraduate courses on irrigation management (Jovanovic and Annandale, 2000b) and crop physiology (Jovanovic et al., 2000b). Many postgraduate students have also been involved with the inclusion of new routines into the model for specific research purposes. For example, the chemical equilibrium routine of Robbins (1991) has been included to enable salt simulations and was used to study the feasibility of irrigating crops with gypsiferous mine water (Annandale et al., 1999b; Annandale et al., 2001; Annandale et al., 2002b). Carbon, nitrogen and phosphorus subroutines have also been included to enable the investigation of nutrient dynamics for the determination of responsible municipal sludge application rates and nitrogen and phosphorus pollution at the field scale (Tesfamariam, 2009; Van der Laan, 2009; Van der Laan et al., 2010).

Future Focus and Opportunities

The emerging story of apparently successful research projects but only patchy adoption of its outputs by the industry remains the key challenge for the irrigation research community. Increases in population and wealth will only make the situation more urgent. On a global scale, it has been estimated that food production needs to double in the next 40 years (Godfray et al., 2010). This directly translates into a need for more irrigation water and more efficient irrigation practices. Already, water-balance studies in 2000 showed that 10 of the 19 water management areas investigated in South Africa could not meet demand (DWAF, 2004). In addition there remains much uncertainty on what the influence of changing weather patterns and elevated CO2 will be on crop-growth processes and water availability, but it is generally agreed that projected climate-change scenarios will lead to additional challenges to the irrigation industry. Prior to 2008, the cost of electricity in South Africa was rated amongst the cheapest in the world (Jumman, 2009). Increases in population and fossil-fuel prices, coupled with infrastructure difficulties, has meant that the country's energy supplier has struggled to meet demand and electricity prices have escalated dramatically. Additional increases of 25% in 2011 and another 25% in 2012 have already been approved, and are expected to further impact negatively on the profitability of irrigation farming Various cost-saving options are now available to irrigators, for example the 'Nightsave' and 'Ruraflex' options, which are being introduced to encourage electricity use for irrigation during off-peak periods (Jumman, 2009). We envisage 4 responses to the above challenges. First, continued advances in existing technology related to the direct measurement of volumetric soil-water content coupled with advances in remote data access through radio communication or via cellular networks, may still develop a cost-effective, real-time system of irrigation. This scenario invokes the concept of 'embodied technology' whereby the data collection and processing is done automatically, and the user receives a daily recommendation without having to understand how any of the technology actually works. Second, new and emerging technology may yet overcome the barriers that have plagued existing technology. For example, space-borne remote sensing can be used to provide regular hydrological information for large areas and though used for research purposes in the past (Bastiaanssen et al. 2000), it is now receiving increased application in irrigationscheduling recommendations (e.g. Santos et al., 2008). Remote sensing is already being used in the Western Cape to assist table-grape and wine farmers to better manage scarce irrigation water resources and nitrogen fertiliser in a project sponsored by the Western Cape Department of Agriculture (see www.GrapeLook.co.za) (Posthumus, 2011). A second project using this technology in irrigated sugarcane production in the Mpumalanga Province, South Africa, has been launched and is co-funded by the WRC and the Department of Agriculture, Forestry and Fisheries. A third project is being planned for irrigated grain crops in the Middle-Orange River catchment (Van Vuuren, 2011). Such new and emerging technologies may, of course, also be very simple. For both highly sophisticated and simple technologies, proper field testing and validation under different environmental and socio-economic conditions should be a non-negotiable requirement. Third, there are approaches that make 'user-friendliness' the main focus. The MyCanesim system described above is one such example. There has been a trend to provide real-time and pre-programmed irrigation scheduling advice via the internet [e.g. Watertight (California State University, USA); Citrus MicroSprinkler Irrigation Scheduler (University of Florida, USA); PlanteInfo (Denmark)]. Many of these systems allow the use of text-messaging technology to deliver the results. Car et al. (2010) investigated the feasibility of using a photograph, taken with a cell-phone and sent to an internet server via MMS technology, to calculate canopy cover and to estimate a crop coefficient (*Kc*). A fourth approach is to use simple tools and engage irrigators in a process of adaptive learning by combining the best of different methods. For example, various strands of information, such as the experiential knowledge of the irrigator, the crop factor, depth of wetting, salt build-up and soil tension are combined to give a more complete picture of the irrigation dilemma (see for example http://thescientistsgarden.blogspot.com). This approach recognises that farmers are intuitively adaptive managers and the use of a robust conceptualisation combined with simple monitoring presents a way to structure their learning (Stirzaker, 2011). Whichever approach we follow, biophysical scientists will need to remember the words of caution of their social science colleagues who have pointed out that technology-driven approaches often fail to capture the specific goals of the farmers and to understand the constraints under which they operate (Vanclay, 2003) and that adoption is driven by the relative advantage over current practice, risk reduction and the compatibility with existing farm practices (Pannell et al., 2006).

CONCLUSIONS

As an important food producer and the largest user of freshwater resources in South Africa, the irrigation industry has a vital role to play in achieving the country's water goal of 'Some, for all, forever'. WRC-funded research efforts over the past 4 decades to develop, improve and promote the use of irrigation-scheduling tools in South Africa have been impressive, but challenges remain and much work still needs to be done, especially to support the application of such tools. No single method of irrigation scheduling has met with universal appeal. In most cases new developments have only taken root where long and consistent back-up was provided, either by the scientists involved themselves or by consultants. There is no simple way to measure water-use efficiency, but when the effort is made, the results invariably show that there is vast room for improvement. It is frustrating that uptake of novel technologies has been so slow. This is not unique to South Africa, and begs the question of whether or not the approach by biophysicists has been somewhat naive? Irrigation is just one part of a profitable farm business, and researchers have perhaps often not always fully understood the constraints under which irrigators operate. Ultimately, together as water users, we will need to learn our way into the future and become better managers of water and the solutes it carries; this will involve a whole range of different technologies. We propose 4 responses to these challenges worth pursuing. Firstly, continue to advance existing technology related to direct measurement of volumetric soil-water content coupled with advances in remote data access through radio communication or via cellular networks; secondly, new and emerging technology may yet overcome the barriers that have plagued existing technology; thirdly, approaches that make 'user-friendliness' the main focus should receive attention; and finally, we need to develop simple monitoring tools and engage irrigators in a process of adaptive learning by combining the best that different methods have to offer. Given the urgency of the need to save water, and the sound understanding of the physical principles governing the functioning of the soil-plant-atmosphere system developed by science over the years, more determined effort is needed to bridge the gap between science and the application of this knowledge on farm.

ACKNOWLEDGEMENTS

The authors would like to thank the WRC for continued support over many years, and for dragging us out of our technical 'comfort zones' and encouraging us to cross the chasm – from theory to practice.

REFERENCES

1. Allen RG, Pereira LS, Raes D and Smith M (1998) Crop evapotranspiration. FAO Irrigation and Drainage Paper 56. Food and Agriculture Organization (FAO), Rome, Italy. 300 pp.
2. Annandale JG, Beletse Yg, DE Jager PC, Jovanovic NZ, Steyn JM, Benadé N, Lorentz SA, Hodgson FDI, Usher B, Vermeulen D and AKEN ME (2007) Predicting the Environmental Impact and Sustainability of Irrigation with Coal Mine Water. WRC Report No. 1149/1/07, Water Research Commission, Pretoria, South Africa.175 pp.
3. Annandale JG, Benadé N, Jovanovic NZ, Steyn JM and DU Sautoy (1999a) Facilitating Irrigation Scheduling by Means of the Soil Water Balance Model. WRC Report No. 753/1/99, Water Research Commission, Pretoria, South Africa. 285 pp.
4. Annandale JG, Jovanovic NZ, Benadé N and Tanner PD (1999b) Modelling the long-term effect of irrigation with gypsiferous water on soil and water resources. *Agric. Ecosyst. Environ.* 76 109-119.
5. Annandale JG, Campbell GS, Olivier FC and Jovanovic NZ (2000) Predicting crop water uptake under full and deficit irrigation. An example using pea (*Pisum sativum* cv. Puget). *Irrig. Sci.* 19 65-72.
6. Annandale JG, Jovanovic NZ, Mpandeli NS and Lobit P (2002a) Two Dimensional Energy Interception and Water Balance Model

for Hedgerow Tree Crops. WRC Report No. 945/1/02 Water Research Commission, Pretoria, South Africa. 217 pp.

7. Annandale JG, Jovanovic NZ, Tanner PD, Benadé N, DU Plessis HM (2002b) Sustainability of irrigation with gypsiferous mine water. *Mine Water Environ.* 21 81-90.
8. Annandale JG, Jovanovic NZ, Pretorius JJB, Lorentz SA, Rethman NFG and Tanner PD (2001) Gypsiferous mine water use in irrigation on rehabilitated open cast mine land: crop production, soil water and salt balance. *Ecol. Eng.* 17 153-164.
9. Annandale JG and Stockle CO (1994) Fluctuation of crop evapotranspiration coefficients with weather: a sensitivity analysis. *Irrig. Sci.* 15 1-7.
10. Annandale JG, Steyn JM, Benadé N, Jovanovic NZ and Soundy P (2005) Technology Transfer of the Soil Water Balance (SWB) Model as a User-Friendly Irrigation Scheduling Tool. WRC Report No. TT 251/05, Water Research Commission, Pretoria, South Africa. 80 pp.
11. Annandale JG, Van der Westhuizen AJ and Olivier FC (1996) Die Fasilitering van Tegnologie Oordrag deur Verbeterde Besproeiingsriglyne vir Groente en 'n Meganistiese Gewasmodeleringbenadering (Facilitating Technology Transfer through Improved Irrigation Guidelines of Vegetables and a Mechanistic Crop Modelling Approach). WRC Report No. 476/1/96, Water Research Commission, Pretoria, South Africa. 179 pp. ARC-ISCW (2005) Overview of the Agricultural Natural Resources of South Africa.
12. Arc-iscw Report No.GW/A/2004/13. (CD-ROM) ARC-Institute for Soil, Climate and Water, Pretoria, South Africa. 129 pp.
13. Bastiaanssen WGM, Molden DJ and Makin IW (2000) remote sensing for irrigated agriculture: examples from research and possible applications. *Agric. Water Manage.* 46 137-155.
14. Bembridge TJ (2000) Guidelines for Rehabilitation of Small-Scale Farmer Irrigation Schemes in South Africa. WRC Report No. 891/1/00, Water Research Commission, Pretoria, South Africa. 199 pp.
15. Benadé N, Annandale JG, Jovanovic NZ, Meiring JA and Crous CR (2002) The Development of An Integrated Information System for

Irrigation Water Management using the WAS SWB and RiskMan Computer Models. WRC Report No. 946/4/02, Water Research Commission, Pretoria, South Africa. 282 pp. BENNIE ATP (1995) Sound water management concepts and their application at farm level. *Proc. Southern African Irrigation Symposium*. 4-6 June 1991, Durban, South Africa. WRC Report No. TT 71/95 Water Research Commission, Pretoria, South Africa. pp. 287-294.

16. Bennie ATP, Coetzee MJ, Van Antwerpen R, Van Rensburg LD and Burger RD (1988) Water Balance Model for Irrigation Based on Soil Profile Water Supply Rate. WRC Report No. 144/1/88. Water Research Commission, Pretoria, South Africa. 405 pp.
17. Bennie ATP, Strydom MG and Very HS (1998) Use of Computer Models for Agricultural Water Management on Ecotope Level. WRC Report No. TT 102/98. Water Research Commission, Pretoria, South Africa. 250 pp.
18. Bennie ATP, Van Rensburg LD, Strydom MG and DU Preez CC (1997) Reaksie van Gewasse op Voorafgeprogrammeerde Tekortbesproeiing (Reaction of Crops to Pre-Programmed Deficit Irrigation). WRC Report No. 423/1/97. Water Research Commission, Pretoria, South Africa. 225 pp.
19. Bennie ATP and Hensley M (2001) Maximizing precipitation in dryland agriculture in South Africa – a review. *J. Hydrol.* 241 124-139.
20. Beukes O, Volschenk T, DE Villiers JF and Karsten JHM (2003) Deficit Irrigation Studies to Improve Irrigation Scheduling in Deciduous Fruit Orchards. WRC Report No. 892/2/03. Water Research Commission, Pretoria, South Africa. 100 pp.
21. Boedt LJJ and Laker MC (1985) the Development of Profile Available Water Capacity Models. WRC Report No. 98/1/85. Water Research Commission, Pretoria, South Africa. 448 pp.
22. Botha CAJ, Steyn GJ and Stevens JB (2000) Factors which Influence the Acceptability of Irrigation Scheduling with Specific Reference to Scheduling Models. WRC Report No. 893/1/00. Water Research Commission, Pretoria, South Africa. 199 pp.
23. Botha FJP, Bennie ATP and Burger RduT (1983) Water Use Efficiency of Irrigated Crops as Influenced by Varying Cultivation Practices and Root Configurations. Department of Soil Science, UOFS, Bloemfontein. 391 pp.

24. Buchheim JF and Ploss LF (1977) Computerized irrigation scheduling using neutron probes. *Proc. 1977 Annual Conference of the American Society of Agricultural Engineers*. ASAE, St. Joseph, Michigan, US.
25. Burgers MS and KIRK RD (1993) Practical Scheduling of Irrigation in the Northern Transvaal. WRC Report No. 152/1/93. Water Research Commission, Pretoria, South Africa. 161 pp.
26. Car N, Christen E, Hornbuckle J And Moore G (2010) ET-based irrigation DSS using mobile phones where remote sensing is not applicable. *Proc. Irrigation Australia Ltd Conference*, 8-10 June 2010, Sydney, Australia.
27. Charlesworth PB (2005) Irrigation Insights No. 1 - Soil Water Monitoring (2nd edition). National Program for Irrigation Research and Development, CSIRO Publishing, Melbourne, Australia.
28. Costa JM, Ortuno MF And Chaves MM (2007) Deficit irrigation as a strategy to save water: Physiology and potential application to horticulture. *J. Integrated Plant Biol.* 49 1421-1434.
29. Crosby C (2004) The GWK limited irrigation scheduling service. *The Water Wheel* May/June 2004. 14-17.
30. Crosby CT and Crosby CP (1999) Sapwat: A Computer Program for Establishing Irrigation Requirements and Scheduling Strategies in South Africa. WRC Report No. 624/1/99. Water Research Commission, Pretoria, South Africa. 12 pp.
31. Crosby CT, DE Lange M, Stimie CM, Van DER Stoep I (2000) A Review of Planning and Design Procedures Applicable to Small-Scale Farmer Irrigation Projects. WRC Report No. 578/2/00. Water Research Commission, Pretoria, South Africa. 245 pp.
32. DE Jager JM and Kennedy JA (1996) Weather-Based Irrigation Scheduling for Various Farms (Commercial and Small-Scale). Water Reports 8: 33-38. Food and Agriculture Organization (FAO), Rome, Italy.
33. DE Jager JM and Mottram R (1996) saving water by scheduling irrigation using weather data. In: Sustainability of Irrigated Agriculture - the Future of Irrigation under Increased Demand from Competitive Uses of Water and Greater Needs for Food Supply (special session). Volume 1-E. *Transactions 16th International Congress on Irrigation and Drainage*. Cairo, Egypt. 153-166.
34. DE Jager JM, Mottram R and Kennedy JA (2001) Research on

a computerised weather based irrigation water management system. WRC Report No. 581/1/01. Water Research Commission, Pretoria, South Africa. 180 pp.

35. DE Jager JM and Van ZYL WH (1989) Atmospheric evaporative demand and evaporation coefficient. *Water SA* 15 103-110.

36. DE Jager JM and Singels A (1994) Research report by the Irrigation Planning Unit to the Department of Agriculture, Pretoria, South Africa. 6 pp.

37. DE Jager JM, Van ZYL WH, Kelbe BE and Singels A (1987) Research on a weather service for scheduling the irrigation of winter wheat in the OFS. WRC Report No. 117/1/87. Water Research Commission, Pretoria, South Africa. 281 pp.

38. Denmead OT and Shaw RH (1962) Availability of soil water to plants as affected by soil moisture content and meteorological conditions. *Agron. J.* 54 385-90.

39. Doorenbos J and Pruitt WO (1977) Guidelines for predicting crop water requirements. FAO Irrigation and Drainage Paper 33 Food and Agriculture Organization (FAO), Rome, Italy. 193 pp.

40. Dwaf (Department Of Water Affairs and Forestry, South Africa) (2004) *National Water Resource Strategy.* (1st edn.). September 2004. Department of Water Affairs and Forestry, Pretoria, South Africa. 150 pp.

41. English MJ and Nakamura BC (1982) Deficit irrigation: Effects of irrigation frequency on crop yield. *Proc. Technical Conf. on Water, Energy and Economic Alternatives.* The Irrigation Association, Falls Church, Virginia, US. 162-172.

42. Fanadzo M, Chiduza C, Mnkeni PNS, Stoep I and Stevens J (2010) Crop production management practices as a cause for low water productivity at Zanyokwe Irrigation Scheme. *Water SA* 36 27-36.

43. FAO (2005) *South Africa: Water and Food Security Country Profiles.* URL: http://www.fao.org/nr/water/aquastat/countries/south_africa/south_africa_cp.pdf

44. Fessehazion, MK, Stirzaker RJ, Annandale JG, And Everson CS (2011) Improving nitrogen and irrigation water use efficiency through adaptive management: A case study using annual ryegrass. *Agric. Ecosyst. Environ.* 141 350-358.

45. Fischer HH (1995) Scheduling of deficit irrigation for improved

water use efficiency. *Proc. Southern African Irrigation Symposium,* 4-6 June 1991, Durban, South Africa. WRC Report No. TT 71/95 Water Research Commission, Pretoria, South Africa. pp. 344-349.

46. Fischer HH and Nel PC (1990) Drupbespoeiing by Tamaties (Drip Irrigation in Tomatoes). WRC Report No. 185/1/90. Water Research Commission, Pretoria, South Africa. 181 pp.
47. Gardner WR (1960) Dynamic aspects of water availability to plants. *Soil Sci.* 89 63-73.
48. Gardner W and Kirkham D (1952) Determination of soil moisture by neutron scattering. *Soil Sci.* 73 391.
49. Godfray HC, Beddington JR, Crute IR, Haddad L, Lawrence D, Muir JF, Pretty J, Robinson S, Thomas SM And Toulmin C (2010) Food security: the challenge of feeding 9 million people. *Science* 327 812-818.
50. Grant S (2000) Five-step irrigation schedule: Promoting fruit quality and vine health. *Winegrowing* May/June 2000, 1-4.
51. Green GC (1985) Estimated Irrigation Requirements of Crops in South Africa, Part 1. Memoirs on the Agricultural Resources of South Africa, No. 2. Department of Agriculture and Water Supply. Pretoria, South Africa.
52. Haarhoff D (2011) Personal communication via e-mail on 21 October 2011. Manager: Research and Development, Griekwaland- Wes Koöperasie, Douglas, South Africa.
53. Hensley M and DE Jager JM (1982) the Determination of the Profile Available Water Capacities of Soils. Department of Soil Science, University of Fort Hare, Alice, South Africa. 282 pp.
54. Idso SB, Jackson RD, Pinter PJ, Reginato RJ and Hatfield JL (1981) Normalizing the stress-degree-day parameter for environmental variability. *Agric. Meteorol.* 24 45-55.
55. Jackson RD, Idso SB, Reginato RJ and Pinter PJ (1981) Canopy temperature as a crop water stress indicator. *Water Resour. Res.* 17 1133-1138.
56. Jovanovic NZ and Annandale JG (1999) An FAO crop factor modification to SWB makes inclusion of crops with limited data possible: Examples for vegetable crops. *Water SA* 25 181-190.

57. Jovanovic NZ, Annandale JG And Mhlauli NC (1999) Field water balance and SWB parameter determination of six winter vegetable species. *Water SA* 25 191-196.

58. Jovanovic NZ and Annandale JG (2000a) Crop growth model parameters of 19 summer vegetable cultivars for use in mechanistic irrigation scheduling models. *Water SA* 26 67-76.

59. Jovanovic NZ and Annandale JG (2000b) SWB (Soil Water Balance): A computer tool for teaching future irrigation managers. *J. Nat. Resour. Life Sci. Educ.* 29 15-22.

60. Jovanovic NZ, Annandale JG, Claassens AS, Lorentz SA, Tanner PD, Aken ME and Hodgson FDI (2002) Commercial production of crops irrigated with gypsiferous mine water. *Water SA* 28 413-421.

61. Jovanovic NZ, Annandale JG and Hammes PS (2000a) Teaching crop physiology with the soil water balance (SWB) model. *J. Nat. Resour. Life Sci. Educ.* 29 23-30.

62. Jovanovic NZ, Annandale JG and Nel AA (2000b) Calibration and validation of the SWB model for sunflower (*Helianthus annuus L.*). *S. Afr. J. Plant Soil* 17 117-123.

63. Jumman A (2009) A Framework to Improve Irrigation Design and Operating Strategies in the South African Sugarcane Industry. M.Sc. (Eng) Dissertation, University of KwaZulu-Natal, Pietermaritzburg, South Africa. 15 pp.

64. Laker MC (1982) Appendix 8.1 A provisional simple model for predicting PAWC for certain Ciskeian soils. In: Hensley M and De Jager JM (ed.) *the Determination of the Profile Available Water Capacities of Soils*. Department of Soil Science, University of Fort Hare, Alice, South Africa. pp. 235-254

65. LAKER MC (1985) The utilization of heavier soils for irrigation and using the water storage capacities of soils to reduce design peaks. *Proc. SABI Conf., Christiana,* 14-17 March 1985, Christiana, South Africa. Available from mlaker@telkomsa.net.

66. LAKER MC, BOEDT LJJ and HENSLEY M (1987) Pre-dawn leaf water potential as indicator of plant water stress – with special reference to problems encountered under conditions of high evaporative demand. *Proc. Int. Conf. on Measurement of Soil and Plant Water Status* Vol. 2, 6-10 July 1987, Utah State Univ., Logan, US. pp. 25-29.

67. Laker MS (2004) the Effect of Atmospheric and Soil Conditions on the Grapevine Water Status. M.Sc. Agric. dissertation. University of Stellenbosch, Stellenbosch. 107 pp.

68. Lategan E and Howell CL (2010) the partial rootzone drying (PRD) of merlot in the Breede River Valley (Part 1): Irrigation volumes, plant water stress and vigour. *Wynboer Technol. Yearbook 2010*, pp. 19-21.

69. Leib BG, Hattendorf M, Elliot T And Matthews G (2002) Adoption and adaption of scientific irrigation scheduling: trends from Washington, USA as of 1998. *Agric. Water Manage*. 55 105-120.

70. Marais D, Rethman NFG and Annandale JG (2002) Water Use and Water Use Efficiency of Fodder Crops under Irrigation: Part 1 – Annual Subtropical Crops. WRC Report No. 573/1/02. Water Research Commission, Pretoria, South Africa. 76 pp.

71. Meyer WS, Oosterhuis DM, Berliner PR, Green GC and Van Der Merwe AJ (1987) Evapotranspiration and Water Use Studies in Wheat and Soybeans with the Help of the Weighing Lysimeter Technique. WRC Report No. H2/1/87. Water Research Commission, Pretoria, South Africa. 4 pp.

72. Miller DE (1977) Deficit high frequency irrigation of sugarbeet, wheat and beans. *Proc. Speciality Conf. on Water, Irrigation and Management for Drainage*. ASCE, New York.

73. Mkhize SS, Vanassche FMG and Laker MC (1996) the Response of Citrus Seedlings to Soil Compaction and Variations in Soil Water Potential in the Upper Range of Plant Available Water. WRC Report No. 261/1/96. Water Research Commission, Pretoria, South Africa. 250 pp.

74. Monteith JL (1977) Climate and efficiency of crop production in Britain. *Philos. Trans. R. Soc. London*, Ser. B 281 277-294.

75. Mottram R and De Jager JM (1994) Research on maximizing irrigation project efficiency in different soil-climate-irrigation situations. WRC Report No. 226/1/94. Water Research Commission, Pretoria, South Africa.197 pp.

76. Myburgh PA (2011a) Response of *Vitis vinifera* L. cv. Merlot to low frequency drip irrigation and partial root zone drying in the Western Cape coastal region. *S. Afr. J. Enol. Vitic*. 32 89-103.

77. Myburgh PA (2011b) Personal communication via e-mail on

20 October 2011. Soil scientist, ARC-Infruitec/Nietvoorbij, Stellenbosch. NEL AA (1995a) Irrigation timing using leaf water potential of wheat. *Proc. Southern African Irrigation Symposium*. 4-6 June 1991, Durban, South Africa. WRC Report No. TT 71/95. Water Research Commission, Pretoria, South Africa. pp. 336-340.

78. Nel FJ (1995b) Irrigation management on large farms, with special reference to the use of a neutron hydroprobe. *Proc. Southern African Irrigation Symposium,* 4-6 June 1991, Durban, South Africa. WRC Report No. TT 71/95. Water Research Commission, Pretoria, South Africa. 305-307.

79. Nieuwoudt WL, Backeberg GR And DU Plessis Hm (2004) the value of water in the South African economy: Some implications. *Agrekon* 43 162-182.

80. Olivier FC, Lecler NL and Singels A (2009) Increasing Water Use Efficiency of Irrigated Sugarcane by Means of Specific Agronomic Practices. WRC Report No. 1577/1/09. Water Research Commission, Pretoria, South Africa. 127 pp.

81. Pannell DJ, Marshal GR, Barr N, Curtis A, Vanclay F and Wilkinson R (2006) Understanding and promoting adoption of conservation practices by rural landholders. *Aust. J. Exp. Agric.* 46 1407-1424.

82. Passioura JB (1988) Water transport in and to roots. *Ann. Rev. Plant Phys.* 39 245-265.

83. Pavel EW and DE Villiers AJ (2004) Responses of mango trees to reduced irrigation regimes. *Acta Hortic.* 646 63-68.

84. Pavel EW, Vanassche FMG and Grossman YL (2003) Optimisation of Irrigation Management in Mango Trees by Determination of Water and Carbon Demands to Improve Water Use Efficiency and Fruit Quality. WRC Report No. 1136/1/03. Water Research Commission, Pretoria, South Africa. 87 pp.

85. Posthumus C (2011) the intelligent pixel – helping the farmer. *SABI Magazine* April/May 2011, pp 10-11.

86. Ratliff LF, Ritchie JT and Cassel DK (1983) Field measured limits of soil water availability as related to laboratory-measured properties. *Soil Sci. Soc. Am. J.* 47 770-775.

87. Reginato RJ (1995) Crop water stress assessment for irrigation management. *Proc. Southern African Irrigation Symposium*. 4-6

June 1991, Durban, South Africa. WRC Report No. TT 71/95. Water Research Commission, Pretoria, South Africa. pp. 284-286.

88. Reinders FB, Van Der Stoep I, Lecler NL, Greaves KR, Vahrmeijer Jt, Benadé N, DU Plessis FJ, Van Heerden PS, Steyn JM, Grové B, Jumman A and Ascough G (2010) Standards and Guidelines for Improved Efficiency of Irrigation Water Use from Dam Wall Release to Root Zone Application: Guidelines. WRC Report No. TT 466/10. Water Research Commission, Pretoria, South Africa. 234 pp.
89. Ichards LA and Neal OR (1936) some field observations with tensiometers. *Soil Sci. Soc. Am. Proc.* 1 71.
90. Robbins CH (1991) Solute transport and reactions in salt-affected soils. In: Hanks RJ and Ritchie JT (eds.) *Modelling Plant and Soil Systems. Agron. Mono.* No. 31. ASA-CSSA-SSSA, Madison, WI. 365-395.
91. Santos C, Lorite IJ, Tasumi M, Allen RG and Fereres E (2008) Integrating satellite-based evapotranspiration with simulation models for irrigation management at the scheme level. *Irrig. Sci.* 26 277-288.
92. Savage MJ and Wiebe HH (1987) Effect of conductance on water vapor pressure determination of leaf water potential. *Proc. Int. Conf. on Measurement of Soil and Plant Water Status* Vol. 2. 6-10 July 1987, Utah State Univ., Logan, US. 55-61.
93. Savage MJ, Graham and Lightbody KE (2000) An Investigation of the Stem Steady State Heat Energy Balance Technique in Determining Water Use by Trees. WRC Report No. 348/1/00. Water Research Commission, Pretoria, South Africa. 181 pp.
94. Singels A (2007) A new approach to implementing computer-based decision support for sugarcane farmers and extension staff. The case of My Canesim. *Proc. Int. Soc. Sugar Cane Technol.* 26 211-219 (also published in *Sugar Cane Int.* 26 22-25).
95. Singels A and Smith MT (2006) Provision of irrigation scheduling advice to small-scale sugarcane farmers using a web based crop model and cellular technology: A South African case study. *Irrig. Drain.* 55 363-372.

96. Singels A and Smith MT (2009) Real Time Irrigation Advice for Small-Scale Sugarcane Production Using A Crop Model. WRC Report No. K5/1576/4. Water Research Commission, Pretoria, South Africa. 147 pp.

97. Smith M, Allen RG and Pereira LS (1996) Revised FAO methodology for crop water requirements. *Proc. of the Int. Conf. on Evapotranspiration and Irrigation Scheduling,* San Antonio, Texas, USA. 133-140.

98. Spreer W, Nagle M, Neidhart S, Carle R, Ongprasert S And Müller J (2007) Effect of regulated deficit irrigation and partial rootzone drying on the quality of mango fruits (*Mangifera indica* L., cv. 'Chock Anan'). *Agric. Water Manage.* 88 173-180.

99. Stevens JB, Duvel GH, Steyn GJ and Marobane W (2005) The Range, Distribution and Implementation of Irrigation Scheduling Models and Methods in South Africa. WRC Report No. 1137/1/05. Water Research Commission, Pretoria, South Africa. 208 pp.

100. Stevens JB and Stirzaker RJ (2010) Wetting Front Detector Transfer of Technology. WRC Report No. K8/599/4. Water Research Commission, Pretoria, South Africa. 30 pp.

101. Steynberg RE, Nel PC and Rethman NFG (1994) Waterverbruik en Waterverbruiksdoeltreffendheid van Gematigde Aanplante Weidings onder Bespoeiing (Water Use and Water Use Efficiency of Temperate Planted Pastures under Irrigation). WRC Report No. 257/1/94. Water Research Commission, Pretoria, South Africa.

102. Stirzaker RJ (1999) The problem of irrigated horticulture: matching the biophysical efficiency with the economic efficiency. *Agrofor. Syst.* 45 187-202.

103. Stirzaker RJ (2003) When to turn the water off: scheduling microirrigation with a wetting front detector. *Irrig. Sci.* 22 177-185.

104. Stirzaker RJ (2006) Soil Water Monitoring. State of Play and Barriers to Adoption, Irrigation Matters Series 01/06. CRC for Irrigation Futures. Darling Heights, Qld. URL: www.irrigationfutures. org. au/news.asp?catID=12&ID=440

105. Stirzaker RJ (2008) Factors affecting sensitivity of wetting front detectors. In: Goodwin I and O'Connell MG (eds.) *Proc.5th International Symposium on Irrigation of Horticultural Crops. Acta Hort.* 792 647-654.

106. Stirzaker RJ (2011) Strategy, tactics and heuristics for managing solutes in horticultural crops. *Acta Hort.* 889 59-65. STIRZAKER RJ and HUTCHINSON PA (2005) Irrigation controlled by a wetting front detector: field evaluation under sprinkler irrigation. *Aust. J. Soil Res.* 43 935-943.

107. Stirzaker R, Stevens J, Annandale J, Maeko T, Steyn J, Mpandeli S, Maurobane W, Nkgapele J And Jovanovic N (2004) Building Capacity in Irrigation Management with Wetting Front Detectors. WRC Report No. TT 230/04. Water Research Commission, Pretoria, South Africa. 85 pp.

108. Stirzaker RJ, Stevens JB, Annandale JG and Steyn JM (2010a) Stages in the adoption of a wetting front detector. *Irrig. Drain.* 59 367-376

109. Stirzaker RJ, Steyn JM, Annandale JG, Adhanom G, Van Der Laan M and M'marete C (2010b) Adapting the Wetting Front Detector to Small-Scale Furrow Irrigators and Providing a Basis for the Interpretation of Salt and Nutrient Measurements from the Water Sample. WRC Report No. 1574/1/10. Water Research Commission, Pretoria, South Africa. 99 pp.

110. Streutker A (1978) Tensiometer controlled medium frequency topsoil irrigation: a technique to improve agricultural water management. *Water SA* 4 134-155.

111. Streutker A, Molenaar HW, Hamman H, Nel CC and Mulder JH (1981) Besproeiing, Gewasopbrengs en Dreinering op die Vaalhartsbesproeiingskema: 2. Die Voorkoms van Verbrakte Grond en die Invloed van Dreinering Daarop (Irrigation, Crop Yield and Drainage at the Vaalharts Irrigation Scheme: 2. The Occurrence of Salt-Affected Soils and the Effect of Drainage on It). *Water SA* 7 175-184.

112. Strydom MG (1998) Kwantifisering van Insetparameters vir die Besproeiingskeduleringsprogram BEWAB vir Aartappels en Erte. M.Sc. (Agric) Dissertation, University of the Free State, Bloemfontein, South Africa. 207 pp.

113. Tanner CB and Sinclair TR (1983) Efficient water use in crop production: research or re-search? In: Taylor HM, Jordan WR and Sinclair TR (eds.) *Limitations to Efficient Water Use in Crop Production*. American Society of Agronomy, Crop Science Society

of America, and Soil Science Society of America, Madison, Wisconsin. pp 1-27.

114. Tesfamariam EH (2004) Modelling the Soil Water Balance of Canola *Brassica napus* L (Hyola 60). M.Sc. (Agric) Dissertation. University of Pretoria, South Africa. 120 pp.
115. Tesfamariam EH (2009) Sustainable Use of Sewage Sludge as a Source of Nitrogen and Phosphorus in Cropping Systems. Ph.D. Dissertation, University of Pretoria, South Africa. 270 pp.
116. Vanassche FMG and Laker MC (1989) Research on Improving Irrigation Management Based on Soil Water Monitoring and Detailed Knowledge of Profile Available Water Capacities. WRC Report No. 166/1/89. Water Research Commission, Pretoria, South Africa. 347 pp.
117. Van Averbeke W and Netshithuthuni C (2010) Effect of irrigation scheduling on leaf yield of non-heading Chinese cabbage (*Brassica rapa* L. subsp. *chinensis*). *S. Afr. J. Plant Soil* 27 322-326.
118. Vanclay F (2003) Social principles to inform agriculture. In: Wilson BP and Curtis A (eds.) *Agriculture for the Australian Environment. Proc. 2002 Australian Academy of Science Fenner Conference on the Environment*. Charles Sturt University, Australia. 9-24.
119. Van Der Laan M (2009) Development, Testing and Application of a Crop Nitrogen and Phosphorus Model to Investigate Leaching Losses at the Local Scale. Ph.D. Dissertation, University of Pretoria, South Africa. 178 pp.
120. Van Der Laan M, Stirzaker RJ, Annandale JG, Bristow KL And DU Preez CC (2010) Monitoring and modelling draining and resident soil water nitrate concentrations to estimate leaching losses. *Agric. Water Manage.* 97 1779-1786.
121. Van Vuuren L (2011) Improved water use only a satellite away. *The Water Wheel*, September/October 2011. 14-17.
122. Van ZYL WH and DE Jager JM (1992) Effect of climate on plant evaporation coefficients for the potato crop. *Agric. For. Meteorol.* 60 167-179.
123. Van ZYL WH and DE Jager JM (1994) Research on the Climatic Dependence of Evaporation Coefficients. WRC Report No. 260/1/94. Water Research Commission, Pretoria, South Africa. 166 pp.

124. Van ZYL WH, DE Jager JM and Maree CJ (1989) the relationship between daylight evaporation from short vegetation and the USWB Class A pan. *Agric. For. Meteorol.* 46 107-118.

125. Van ZYL WH, DE Jager JM, Maree CJ and Singels A (1990) Canopy surface conductance of unstressed wheat and its weather dependence. *Water SA* 16 99-104.

126. Van ZYL WH and DE Jager JM (1987) Accuracy of the Penman-Monteith equation adjusted for atmospheric stability. *Agric. For. Meteorol.* 41 57-64.

127. Van ZYL WH and DE Jager JM (1992) Effect of climate on plant evaporation coefficients for the potato crop. *Agric. For. Meteorol.* 60 167-179.

128. Volschenk T, DE Villiers JF and Beukes O (2003) The Selection and Calibration of a Model for Irrigation Scheduling of Deciduous Fruit Orchards. WRC Report No. 892/1/03. Water Research Commission, Pretoria, South Africa. 126 pp.

129. Walker S (2000) Irrigation Requirements of Selected Crops under Small-Scale Production: Linking On-Farm and On-Station Research. WRC Report No. 689/1/00. Water Research Commission, Pretoria, South Africa. 146 pp.

130. Water Research Commission (1995) *Proceedings of the Southern African Irrigation Symposium,* 4-6 June 1991, Durban, South Africa. WRC Report No. TT 71/95. Water Research Commission, Pretoria, South Africa. 435 pp.

Chapter 8

Frankincense and Myrrh Resources of Ethiopia: I Distribution, Production, Opportunities for Dryland Development and Research Needs

Mulugeta Lemenih[1] and Demel Teketay[2]

[1]Wondo Genet College of Forestry,
[2]Ethiopian Agricultural Research Organization, Addis Abeba, Ethiopia

ABSTRACT

Boswellia and Commiphora species are economically and ecologically important plant species found mainly in the horn of Africa particularly in Ethiopia, Somalia and Kenya. They are the source of aromatic gum resins, frankincense and myrrh. Frankincense and myrrh have been valued for their sacred and ceremonial uses as well as in medicinal contexts since several millennia. Still today, they are widely used as raw materials in several industries such as pharmacology, food, beverage, flavouring, liqueurs, cosmetics, detergents, creams and perfumery, paints, adhesive and dye manufacturing. Ethiopia is one of tropical African countries with large potentials of frankincense and myrrh resources and has been known as one of the major producers. Nevertheless, little efforts have been made at national level to explore the vegetation resources that provide these valuable products. Consequently, little is written and known about their distribution, potential production, development opportunities and other relevant issues. This paper presents a review that covers several aspects of these valuable resources including the economic and ecological opportunities they may render for accelerated development and ecological conservation in the vast dry lands of the country.

INTRODUCTION

Vast areas of land, ca. 51%, in Ethiopia are arid to semiarid (NCSS, 1993; Le Houérou, 1996; Tamerie Hawando, 1997) with marginal or no agricultural potentials. Including the dry sub-humid area, the total dry landmass of the country amounts 860,000 km^2 (ca. 71%) (Tamerie Hawando, 1997). Nevertheless, many of the indigenous trees and shrubs in these vast arid and semi-arid lowlands hold known or potential promise for the production of economically valuable products, principally oleo gum resins such as gum acacia (gum arabic and gum talha), frankincense, myrrh, hinna, and gum karaya (Stiles, 1988; EFAP, 1993; FAO, 1995; Kuchar, 1995; Mulugeta Lemenih et al., 2003).

Boswellia and Commiphora species are two of the genera in the family Burseraceae that are mainly found in the hot and dry lowlands of north east Africa with a few species in Arabia and India. For thousands of years the aromatic gum resins, frankincense and myrrh, have been harvested from these genera to supply the demands of old civilizations, many faiths and medical demands. Frankincense and myrrh were once regarded as lifeline to spiritual and physical health and considered precious than gold (Groom, 1981; Hillson, 1988; Wilford, 1997). Still today, frankincense and myrrh are widely used in unprocessed form for fragrance in many religious rituals (Coulter, 1987; Farah, 1994; FAO, 1995). Moreover, frankincense and myrrh are phyto-toxically safe raw materials utilized in many of today's industries such as pharmacology and folk medicines (Mielck, 1970; Tyler, 1993; Wichtl and Bisset, 1994; Bruneton, 1995; Budavari, 1996; DAB, 1997; BHP, 1998), food industries (Council of Europe, 1981; Ford et al., 1992; FAO, 1995), flavouring, beverage and liqueurs (FAO, 1995), cosmetics, detergents, creams and perfumery (Fragrance Foundation, 1983; Tucker, 1986; Wahab et al., 1987; Farah, 1994; FAO, 1995; Leung and Foster, 1996), and in paints, adhesive and dye manufacturing (Krishna-Murthy and Shiva, 1977; FAO, 1995; Anonymous, 2001).

Oleo-gum resins have been articles of great historical commerce in the Horn of Africa in general and Ethiopia in particular (Ahmed, 1982; Coulter, 1987; EFAP, 1993). Today they are among the top export articles of Ethiopia where their export alone accounts for about one-half of the total export values (Anonymous, 2001). For instance, in the years 1996/97, 1997/98, and 1998/99 about 1466, 1925 and 1649 tonnes of gum resins were exported from Ethiopia with the total values of US$ 2.5, 2.6 and 2.1 million, respectively (Anonymous, 2001). Ethiopia was the main trader of frankincense and myrrh more than 4000 years ago (Ahmed, 1982), and today Ethiopia and Somalia are the major producers and exporters of frankincense and myrrh in the world market (FAO, 1995; Coulter, 1987; Wahab et al., 1987).

Despite all these facts, little has been done to explore, study and document the vegetation resources that provide these valuable resources in Ethiopia. Thus, information about them is generally scanty (EFAP, 1993; Tilahun Gebre-medhin, 1997). Few of the attempts in forest valuations and forestry concerns of the country have focused much on the humid and dry sub-humid forest resources. Comparatively, little attention has been given to the woodland vegetations in the arid lowlands (EFAP, 1993), although the national economic contributions, in terms of export, of the semiarid and arid vegetations far outweigh the forests of the humid and sub-humid combined. One probable reason for the ignorance could be lack of information about them. Similarly, because of the marginalized environmental conditions, arid and semiarid lowlands have often been considered as lands of no economic opportunity (Stiles, 1988). Nevertheless, the arid and semi-arid lands account for more than half of the country's land area

(NCSS, 1993; Tamerie Hawando, 1997) and 12% of the population. It is, therefore, urgently compelling to seek for viable strategies that will make the vast arid and semiarid land resources optimally contribute to the national economy as well as buffer the economy of the households living in these marginal arid areas.

This paper presents a review made on some aspects of frankincense and myrrh resources tapped from the arid and semi-arid lowland vegetation resources of Ethiopia. The review is expected to contribute valuable information on the potentials of the arid and semi-arid lowland vegetation assets of the country and the golden opportunity they may provide for the challenges to accelerate economic development in these marginal lands.

WHAT ARE FRANKINCENSE AND MYRRH?

Frankincense and myrrh are resinous plant exudates obtained from certain species of the genera Boswellia and Commiphora, respectively. Both genera belong to the family Bureseraceae. Boswellia and Commiphora species that yield frankincense and myrrh are tropical genera, which are dominantly found around the Horn of Africa, and some in Arabian Peninsula and India (Wahab et al., 1987; Farah, 1994; FAO, 1995; Ermias Dagne et al., 1997).

The word frankincense is derived from the old French term "france encens", meaning "pure incense" or literally, "free lighting" (Tucker, 1986). Frankincense, incense, olibanum, resin, and aromatic products, are all used to refer to the products of frankincense tree. As the product is odoriferous, it is often referred to as incense in general terms while technically this product is referred to as gum-resin (Farah, 1994). The term olibanum is dominantly used exchangably with frankincense (Farah, 1994; Tilahun Gebremedhin, 1997).

Frankincense is produced and exported by few countries. Somalia and Ethiopia are the major producers and exporters (Wahab et al., 1987; Coulter, 1987; Farah, 1994; FAO, 1995). In India, resinous gum obtained from Boswellia serrata is also traded as frankincense under the name Indian olibanum but is graded as an inferior quality olibanum at international market (Farah, 1994; FAO, 1995). The frankincense produced from Boswellia sacra species (Somali name 'Moxor') is called Somali type olibanum in the international trade, and the second frankincense from Boswellia frereana Birdw species (Somali name 'Yagcar') is known by the name luban lami in Arabia (Farah, 1994). Two types of frankincense are distinguished from Ethiopia: the Ogaden quality and Eritrean/Tigray type. The Eritrean/Tigray type olibanum is the most widely traded frankincense in the world (Farah, 1994), and is the gum-resin obtained from Boswellia papyrifera (Del.) Hochst. The Ogaden type is gum-resin type produced in the east and south-eastern parts of the country. This type is known by the name Borena and Ogaden type of olibanum in Ethiopia. The specific source species for these latter types of frankincense are not consistent in the literatures found. However, some of the sources indicated that resins from B. rivae (Engl.), B. ogadensis (Vollesen) (Somali name 'Gended'), B. neglecta (S. moore) (Somali name 'Murufur') and B. microphylla (Chior.) (Somali name 'Muqlay') are collected and traded as frankincense in this area (Girmay Fitwi, 2000; Mulugeta Lemenih et al., 2003). Other species that yield resinous products designated as frankincense may also exist in these parts of Ethiopia, and may even include those species known from Somalia, for instance, B. sacra (Vollesen, 1989).

Myrrh is sometimes called bdellium or oppoponax, although distinction is often made between the three depending on the source species. Others also divide myrrh as scented/perfumery myrrh (called 'bissabol' or oppoponax) and medicinal myrrh (called 'heerabol') (Thulin and Claeson, 1991). Myrrh (heerabol) is a typical name for the gum obtained from Commiphora myrrha (Nees) Engl., and is called true myrrh (FAO, 1995). This is probably the myrrh used dominantly in medical context. Oppoponax/scented myrrh is the name given to the myrrh obtained from Commiphora guidottii Chior. (Thulin and Claeson, 1991; Farah, 1994) or C. erythraea (Ehrenb.) Engl. or C. kataf (FAO, 1995). There are still many Commiphora species that yield gums that are sold under the name myrrh or oppoponax such as Commiphora borannsis Vollesen, C. africana (A. Rich.) Engl., C. habessinica (Berg.) Engl. C. madagascariensis, C. kataf, C. erythraea (Ehrenb.) Engl. and others (Groom, 1981; Tucker, 1986; Mulugeta Lemenih et al., 2003). It is not clear whether the myrrh obtained from these latter species are used for fragrance, medicinal purpose or for adulteration (FAO, 1995). Nevertheless, the gum-resin obtained from C. africana called 'hagar' in Somali is know to be collected and traded for medicinal use in Liban

Zone of the Ethiopian Somali National Regional State (ESNRS) (Mulugeta Lemenih et al., 2003). Furthermore, oleo-gum resin collected from C. erythraea /C. holtziana, also called 'hagar' in Somali is exported from Kenya as oppoponax just like bissabol (Provan et al., 1987), although it is different from the true scented myrrh produced from C. guidottii (Thulin and Claeson, 1991). Myrrh and its various associates obtained from different species of Commiphora are technically referred to as gum (Farah, 1994). The word Myrrh assumed to have been derived from the Assyrian word called 'murru' (Tucker, 1986).

In general, the botanical origin of frankincense resins and the name ascribed to their sources has been a subject of much uncertainty and discussion over many years. This is one of the most difficult areas that should be addressed by research in the future. The situation for myrrh and Commiphora species, is even much more complex (FAO, 1995). These mix-up of names, lack of proper assortments or mixing gum-resins of different species into a single one in shipments, and unrecorded parallel trades between countries have had their own root consequences in making identification of source species, quality controls and source countries very difficult. The fact that collectors, traders, enterprise owners and officials know very little about the uses to which oleo-gum resins traded are put by the final consumers have its own contribution to the absence of quality concepts (Farah, 1994).

THE GENERA BOSWELLIA AND COMMI-PHORA IN ETHIOPIA

General Description

Boswellia and Commiphora belong to the family Burseraceae. This family has 17 genera with estimated 560–600 species (Vollesen, 1989; Gachathi, 1997). This family is widespread in all tropical regions and extending into the subtropics. However, only the two genera, Boswellia and Commiphora with 58 species are known to exist in Ethiopia (Vollesen, 1989). Six of the 58 species belongs to the genus Boswellia and the rest to the genus Commiphora (Vollesen, 1989).

The diagnostic features of the family Burseraceae are that they are trees or shrubs with prominent vertical resin ducts in the bark. The leaves, which are compound, are spirally arranged and crowded at twig-tips. The flowers are rather small and are either solitary or inflorescences usually at the twig-ends, regular with parts in threes to fives, bisexual or more often unisexual, the plants often dioecious. The sepals are fused and are either imbricate or valvate, petals and usually in two whorls. The ovary is superior with three to five carpels and two to five locules. The fruit is usually a drupe, sometimes a capsule. Seeds are without endosperm (Gachathi, 1997).

Botanical Description

Genus Boswellia

The genus Boswellia Roxb. Ex Colebr. is composed of 20 or so species extending from Ivory Coast to India and south to N.E. Tanzania and N. Madagascar but most numerous in north east tropical Africa (Kuchar, 1995, Vollesen, 1989, Gachathi, 1997). These are unarmed shrubs or small to medium-sized trees. The outer bark is often peeling in parchmentry flakes, while the inner bark greenish, with watery aromatic resin, which slowly hardens to a resin on exposure; the wood is with milky latex. Leaves are imparipinnate, mostly congested at the end of branches. Flowers are bisexual and are in panicles or racemes. Calyx is 5-lobed. Petals are 5, free, oblong to ovate, and acute. Stamens are 10 and free. Disc is conspicuous, annular, flat or concave. Ovary is trigonous, (2) 3 (4)-locular; style is simple, and stigma is capitate. Fruit is a (2) 3 (4)-locular obtetrahedroid drupe with woody exocarp, dehiscing with the valves dropping and leaving a central (2)3(4)-winged column with the stones attached. Stones are tetrahedroid when young with a thin papery wing formed from outer layer of endocarp, with apical and/ or lateral and/or basal extensions (horns) (Vollesen, 1989; Kuchar, 1995; Gathachi, 1997).

Table 1: Economically valuable frankincense and myrrh producing Boswellia and Commiphora species found in Ethiopia

Species	Common name	Location in Ethiopia	Sources
Bonoeltia sacra Fluckiger	Mohor (Somali)	ESNRS	Groom (1981); Vollesen (1989)
B. neglecta S. Moore	Murufur (Somali)		Vollesen (1989), Mucha? (1995) & Mulugeta Lenient et at., (2003)
B. ogadensis Vollesen	Gended (Somali)		Vollesen (1989), Mulugeta Lemenih et al. (2003), Kuchar (1995)
B. microphylla Chiov.	Mug.lay (Somali)		Vollesen (1989), Mulugeta Lemenih et al. (2003), Kuchar (1995)
B. papyri/era (Del.) Hochst.	Etan (Amharic) (Tigray /Eritrean type)	North and Northwest Ethiopia	Tilahun Gebremedhin (1997), Abeje Eshete (2002), Vollesen (1989)
B. rivae EngL	Etan (Amharic) (Borena type)	Oromiya/ EsNitS	Vollesen (1989), Girmay Fitwi (2000)
Commiphora myrrha (Nees) Engl.	Molmol (Somali)		Mulugeta Lemenih et al. (2003), Farah (1994), Vollesen (1989)
C truncata Engl	Foah (Somali)		
C guidoth Chiov.	Hadi (Somali)		Thulin and Claeson (1991)

C Itabessinica (Berg) EngL			Tucker (1995), Girmay Fitwi (2000)
C sdtimperi (Berg.) EngL			Girmay Fitwi (2000)
C. boranensis Vollesen	Murfur (Somali)		
C. erythraea (Ehrenb.) Engl.	Hagar (Somali)		Thulin and Claeson (1991)
C. africana (A.Rich.) Engl	Hagar (Somali)		Mulugeta Lemenih et al. (2003); Azene Bekele et al (1993)
C. ogadensis Chiov.	Murfur (Somli)		Vollesen (1989)

Genus Commiphora

The genus Commiphora Jacq. comprises of 150–200 species of which 50–52 species are found in Ethiopia. The genus is a very conspicuous and dominant element in the dry bush lands of NE Africa, and a large number of species are endemic in this area. They are trees or shrubs; branches often terminating in spines; outer bark often papery and peeling, inner bark usually greenish, with resinous–usually aromatic–sap, wood with milky latex. Leaves imparipinnate or 1-3-foliolate (rarely simple), often congested at branch ends. Flowers unisexual and dioecious, usually before the leaves appear or with new leaves, in panicles, cymes, racemes, fascicles or solitary. Calyx 4-lobed. Petals 4, free, linear to elliptic, recurved apically. Stamens (4) 8, didynamous, antisepalous usually the longest, vestigial in female flowers. Disc 4- or 8- lobed or annular, usually glabrous. Ovary 2 (3)-locular; style short; stigma capitate (rarely clavate); ovary absent or vestigial in male flowers. Fruit with a single 1(2)-seeded stone; exocarp fleshy or leathery, splitting into 2 or 4 valves; outer layer of endocarp forming a fleshy pseudaril (Vollesen, 1989; Kuchar, 1995; Gachachi, 1997).

Geographical Distribution in Ethiopia

Natural gum and resin producing trees are mainly found in the drier low lying arid to semi-arid lands in Ethiopia at altitudes varying from 200–1500 m above see level. The gum belt runs in a narrow semi-circle around the western, southern and eastern parts of the country close to the borders as well as in Blue Nile gorges on steep rocky slopes or ridges, lava flows or sandy river valleys, and often associated with gypsum out-crops in the southeastern and eastern part of the country (Thulin and Claeson, 1991; Azene Bekele et al., 1993; EFAP, 1993; Farah, 1994; Fitchl and Admasu Adi, 1994; Tilahun Gebremedhin, 1997). They are dominantly found in the Afar, Amhara, Beneshangul-Gumuz, Gambella, Oromiya, ESNRS and Tigray (Table 3).

Estimated and Actual Production of Frankincense and Myrrh

It is evident that substantial quantities of gum and resins are harvested annually with significant economic importance in Ethiopia. However, little reliable quantitative assessments at country or regional levels have been made. Available estimates, nonetheless, prompt that the total area of woody vegetations with oleo-gum resin producing trees is very large (Girmay Fitwi, 2000; Mulugeta Lemenih et al., 2003). Information obtained from some of the attempts made to provide quantitative estimates at regional level is summarized in Table 2, while the actual national export since 1992 is presented in Table 3. Since no inventory has been carried out on natural gum and

resin resources of the country, the information presented here is neither adequate nor accurate. For example, study made in the Liban Zone of the ESNRS alone confirmed that there is a potential production of 2664990.5 quintals of frankincense, and 1862323 quintals of myrrh as well as large quantity of gum arabic and hagar (Mulugeta Lemenih et al., 2003). While the other remaining zones of the ESNRS are also expected to have more or less comparable potentials, the estimate given in Table 2 for the whole region of the ESNRS is far lower than that of the Liban Zone alone.

Tapping/Production Process

Oleo-gum resins including frankincense and myrrh have been collected and traded from Ethiopia since antiquity. Nevertheless, the production processes such as tapping, resin collections, processing, grading and marketing of the products are little documented. Moreover, no attempts have been made so far to introduce improved production techniques (Azene Bekele et al., 1993; EFAP, 1993; Tilahun Gebrmedhin, 1997). Information obtained from few of the studies made in Ethiopia on traditional tapping systems as well as experiences from other gum resin producing countries are presented here.

Generally oleo-gum resins collection is seasonal and most commonly performed during dry seasons (Murthy and Shiva, 1977; Farah, 1994; Tilahun Gebremedhin, 1997; Mulugeta Lemenih et al., 2003). Tapping during rainy seasons is not recommended due to washing off of the gum-resins by the rain and consequent gum-resins' quality deteriorations.

Tapping techniques differ from place to place and, hence, there is no one described method considered as standard or satisfactory method. A common practice for frankincense production involves blazing (wounding) trees by shaving off the bark using sharp instruments. In Tigray, the frankincense trees are blazed at one-meter height above the ground. A tree is tapped at three spots of 15–20 cm interval for the first time on each of the four sides or, if the tree is small only on two sides facing east and west. Second tapping is done 20–25 days after the first tapping. The wound from the first tapping is removed and refreshed during the next visit, and a third tapping is repeated after 20–25 days following the second tapping. At this time, as well, refreshment of the older wounds are made by removing barks from the upper edges of the former blaze (wound) and by carving down to 2 cm of the lower edge. Each tree is tapped and wounds refreshed at an interval of 15 days until the onset of the rainy season, with a tree being tapped 8–12 times per year. In Tigray, a tree is let to rest for a period of 3–5 years for wound healing after a year of gum collection from it. An estimated yield of 1–3 kg of frankincense is collected from a tree per year (Tilahun Gebremedhin, 1997).

Wounding is not universal, however. In some areas such as Liban Zone of the ESNRS, for instance, collection is made just on tears that naturally exude. There are no incisions artificially made on frankincense trees (Mulugeta Lemehin et al., 2003). While, in Somali, tapping is done by incisions on the body of the frankincense trees where the bark can withstand wounding. The depth of incision depends upon the depth of the resin-bearing ducts of the plants. The initial cut is no more than a scratch. From the small scratches the cuts develop to wider and deeper wounds as the tapping cycle proceed. At the height of the collection season, the average depth of the cuts may measure about 2.5 mm or more and about the size of the palm of a hand (Farah, 1994). Apart from other factors such as the conditions of the plants, the number of tapping usually corresponds to the age of the tree. Some of the fecund and large trees may bear about a hundred wounds at a time, while smaller trees that are being taped for the first time may bear no more than four incisions. Tapping is a simulta-neous process that involves collecting the tears and cleansing (refreshing) the wounds by exposing the resin-bearing ducts for further exudation.

Tapping at the right time is very crucial (Farah, 1994). If the trees are not tapped at the right time, the wound may totally heal. Since it takes time to prepare the plants to produce adequate and superior resin, starting the process all over again will cause a great loss. Commiphora trees are always less wounded artificially than frankincense (Farah, 1994). This may be due to their abundance compared to the frankincense tree resources.

In India, traditionally, the trees are blazed all around or leaving a little space. In some places of India, 15 cm wide shaving of the bark round the bole at a height of 60 cm above the ground level is done to a depth of about 1.5 cm. After a month of initial shaving, freshening to a width of 1.5 cm over the initial one is made. In other places, a wound of 20–25 cm in width and 0.5 cm in depth is made all-round the tree at a height of 105–120 cm from the ground level. After about a fortnight, the exudates is collected. Simultaneous with the collection, the wound is refreshed a little upward. This operation is repeated weekly till the commencement of monsoon. Each tree is tapped on a 3 years cycle (Murty and Shiva, 1977). Although the yield from a tree varies according to locality, tree size, season and surface of the wound exposed, in drier areas a well grown tree of about 30 cm in diameter yields about 2-2.5 kg gum-resin in a year (Murty and Shiva, 1977).

Table 2: Frankincense and myrrh production potentials in different regions of Ethiopia

Regional State	Type of oleo-gum resin	Estimated total area coverage (ha)	Estimated po- tential produc- tion (Quintals)	Source
Afar	1. Gum olsbanum	65,000	1. 2,500	Girmay Fitwi (2000)
	2. Myrrh/Gum commiphora		2. 5,000	
Amhara	1. Gum olthanum (Tigray type)	604,382	1. 2,000,000	Abeje Eshete (2002)
	2. Gum Olthanum (Ogaden type)		2. 11, 923	Anonymous (1997).
Beneshangule	Gum olthanum	100,000	55,000	Girmay Fitwi (2000)
ESNRS	Gum olthanwn (Ogaden type)	150,000	25,000	Girmay Fitwi (2000)
	Myrrh/Gum commiphora		45,000	
Gambella	Gum olthanum		nd	Girmay Fitwi (2000)
	Myrrh/Gum commiphora	420,000	nd	
Oromiya	1. Gum olthanum (Borena type)	430,000	1. 60,000	Girmay Fitwi (2000)
	2. Myrrh / Gum commiphora		2. 15,000	
Tigray	Gum olthan um (Tigray type)	940,000	88, 769.5 to 266,308.5	Tilahun Gebremedhin (1997)

Table 3: Actual natural gum production/export by type for the years 1992–1999 in Ethiopia

Type of Gum				Production/export (Quintals)					
		1992	1993	1994	1995	1996	1997	1998	1999
Tigray/ Briterean Type olibanum		18,004	25,266	30,637	NP	NP	11,923	7,178	14,223
Ogaden type olibanum		317	596	343	478	2,627	58	NP	NP
Borena type olibanum		54	251	1,168	2,005	1,777	106	NP	NP
Myrrh		373	291	968	486	743	1,051	853	NP
Oppoponax		135	33	NP	9	53	147	80	122
Gum acacia		92	269	1,098	1,485	1,130	56	582	36
	Total	18,975	26,706	34,214	4,463	6,330	13,341	8,693	15,378

An improved tapping technique was devised and experimented in India on C. wightii (Bhatt et al., 1989). The method involves tapping using 'Mitchie Golledge Knife, and treatment with 400 mg root feeding of active substance of ethephone (growth regulatory chemical). The result showed enhancement by 22 times oleo-gum resin production over that obtained in the untreated control. The technique was indicated to be inexpensive, safe, requires no specialized skills and can be easily taught to the traditional collectors (Bhatt et al., 1989). The gains from this improved technique include: shorter tapping cuts, reduced tapping frequencies, improved survival of the tapped plants and optimum production.

DEVELOPMENT OPPORTUNITIES FOR ARID AND SEMI-ARID LANDS

General Problems of Dry Lands

As stated above, about 620,000 km^2 (ca. 51%) of the landmass in Ethiopia is arid to semiarid (Le Houérou, 1996; Tamerie Hawando, 1997). Including the dry sub-humid area, the total dry landmass of the country amounts 860,000 km^2 (ca. 71%) (Tamerie Hawando, 1997). In about 22% of this landmass, cultivation is not possible even with the early maturing crops (NCSS, 1993; Tamerie Hawando, 1997). In the remaining parts of these drier areas uneven and unpredictable distribution of rainfall has always been a sever constraint for agricultural performance.

Pastoralism, in different forms, has strong tradition in these vast arid and semi-arid areas of the country. Although it is the sensible mode of production in arid and semi-arid environments, pastoralism is not without weak points. First, its low productivity means the pastoral households are limited to subsistent mode of production where little or none is left for local markets (Steen, 1994). Second, it demands extensive area per household, which implies that it can support only a small population. Even at present low population density, the pastoral economy in Ethiopia could not produce all the food a family needs for basic subsistence, that food shortages are often common during normal dry seasons (Farah, 1997). Thirdly, recurring droughts and frequent fodder failures have been threatening the pastoral economy and their overall socio-economic developments. Moreover, the semi-arid and arid lowlands and Rift Valley in Ethiopia have major problems of salinity and alkalinity, where 36% of the country's total land area is indicated to be potentially susceptible to salinity problems (Tamerie Hawando, 1997).

With rising population and livestock sizes, shrinking land resources, advancing desertification as well as gloomy scenarios in connection with global climatic changes, the prognosis for these marginal lands of Ethiopia does not look good at all. Faster interventions are, therefore, imperative to stop and reverse the trend. On the other hand, the question may be how these vast arid and semi-arid lowland resources could be made productive to support people and contribute to the national economy. The woody vegetation resources that are naturally growing in the arid and semiarid lowlands of Ethiopia may provide enormous opportunities for accelerated economic development in these marginal environments (Mulugeta Lemenih et al., 2003).

Economic Opportunities

The gum and resins from Boswellia and Commiphora species are most valued products in economic terms. Frankincense and myrrh are few of the export articles in Ethiopia (Table 3). Simulta-neously, the economic significances they provide at local level are wide (Farah, 1994; Tilahhun Gebremedhin, 1997; Mulugeta Lemenih et al., 2003). Tapping of frankincense and myrrh provides considerable cash income and employment opportunity (Tilahhun Gebremedhin, 1997; Mulugeta Lemenih et al., 2003). It has been shown that in Liban Zone of the ESNRS collection and sale of oleo-gum resins provide annually an average of US$ 80.00 income per household (Mulugeta Lemenih et al., 2003), an income that was indicated to cover one-third of the annual subsistence of a pastoral household. This contribution from oleo-gum resins was three fold greater than the contribution from arable crops in the area. Similarly, a study made in Tigray revealed that tapping of frankincense provides considerable employment opportunity for the local people (Tilahun Gebremedhin, 1997). This economic incentive provided by the oleo-gum resin extraction has wider implications in the overall socio-economic conditions of households living in arid and semi-arid lowlands. It diversifies their economy and potentially minimizes the risks associated with frequent crop and fodder failures as a result of the recurring droughts.

Generally, the economy of pastoralists and agro-pastoralists in arid and semiarid regions is not able to produce all the food their families need for basic subsistence, and, therefore, food shortages are common even during the normal dry seasons. Consequently, households are forced to consume purchased grains at relatively high prices, while selling their livestock at reduced prices since large livestock is taken to market during the dry seasons. One advantage associated with the vegetation resources in this regard is their ability to produce gum resins only during dry seasons when forage and grains are scarce (Chikamai and Gachathi, 1994), and thus, potentially be used as alternative source of income for dry seasons.

Moreover, frankincense and myrrh are known to have wide pharmacological and industrial uses and potentials (Mulugeta Lemenih and Demel Teketay, 2003), which have not yet been exploited in Ethiopia. If these industries may be introduced they will provide enormous employment for the local people and significant contribution to the national economy as well. In addition, the vegetation resources in these marginal arid lowlands have other several economic roles for the local people. Livestock production, which supplies the major needs of the pastoral families, is heavily dependent on the plant biomass for fodder provision (Kuchar, 1995). Particularly in dry season as well as during droughts, the trees and shrubs are the only source of fodder for the livestock. The family of Burseraceae and Mimosaceae are known for their provision of nutritious fodder, and virtually all the Burseraceae are palatable to livestock particularly in late dry-season (Kuchar, 1988; Farah, 1994). The vegetation and their oleo-gum resins also provide other products and services of local economic importance such as medicine, emergency food, etc. Therefore, accelerated economic development in these marginal arid ecosystems lies in recognizing the potentials that exist in the vegetation assets for the various oleo-gum resin resources and incorporating them in the development strategy.

Ecological Significance

Desertification, which is defined as land degradation in arid, semi-arid, and dry sub-humid areas (Le Houérou, 1996), is reported to have threatened about 72% of Ethiopia's landmass (Tamerie Hawando, 1997). On the other hand, arid and semi-arid lands are generally very vulnerable and easily irreversible once degraded (Eden and Parry, 1996). Many techniques of desertification controls have been in trial in different parts of the World. These methods dominantly involve the redressing of the vegetation covers and control of soil-wind erosion. Although control of the soil-wind erosion that results from vegetation destruction is a first step in desertification control, strategies that combine economics and ecology are needed (Shengyue and Lihua, 2001). Desertifica-tion is derived by factors related to the rural economy. Thus, its control needs to relate to the development of the rural economy in order to form an eco-economic model that can be applied for control. Techniques that support the sustainable production system and promote the economies of the rural community to lead to economic and ecological betterment are required (Shengyue and Lihua, 2001).

The gum and resins producing plants are found in the hot and dry arid and semi-arid areas that are prone to desertification. Consequently, they dispose several ecological significances, while possessing huge economic potentials to buffer and improve the economy of the communities living there. These plants have proved useful as windbreaks and shelter belts against desert encroachment and hence desertification, their canopies intercept rain drops while root systems are effective in reducing soil erosion, thereby stabilizing soils (Mugah et al., 1997). The non-destructive extraction of oleo-gum resins means, therefore, conservation of the vegetation resources and the protection of the vast marginal arid and semi-arid ecosystems. Consequently, the environ-mental benefits of these plant resources in the arid and semi-arid regions are significant.

RESEARCH NEEDS

Despite the early recognition, classification and nomenclature of members of the two genera, Boswellia and Commiphora in tropical Africa have remained unstable (Gachathi, 1997). They have been described as taxonomically difficult, frustrating or simply confusing (Vollesen, 1989; Gachathi, 1997). In general several aspects need research. The following are some of the topics. Although it is accepted that frankincense and myrrh are produced from plants of the family Burseraceae and the genera Boswellia and Commiphora, respectively, there are mix-up and clashes of names, which make identification of which species is the real source of which incense and myrrh. The botanical origins of the frankincense and myrrh and the names ascribed to their sources are generally uncertain (Thulin and Claeson, 1991; FAO, 1995). Moreover, little appears to be definitely known on the source country itself (Farah, 1994; FAO, 1995). For instance, while large quantity of the myrrh traded from Somalia is known to originate from east and southeast Ethiopia (Coulter, 1987; Farah, 1994; FAO, 1995), Ethiopia is not often mentioned as a major producer of myrrh as does Somalia in many references (e.g., Thulin and Claeson, 1991). Furthermore, the olibanum in the Middle East markets is said by some sources to come principally from three Boswellia species namely B. carteri and B. frereana in Somalia and B. sacra in southern Arabia. However, Thulin and Claeson (1991) concluded that B. carteri is simply a variable form of B. sacra and should not be afforded a separate species status. Similarly, while Ethiopia is producing frankincense in the east and southeastern part known by the title Ogaden quality the source species are not known. Moreover, it is uncertain whether the Ogaden quality olibanum is of the same quality, lower or higher compared to the Somali Olibanum. Likewise, Commiphora species known to yield gum-resins identified as myrrh are numerous and include: C. myrrha, C. guidottii, C. serrulata, C. erythraea, etc. Some sources indicate that the true myrrh, for instance, is originated from C. molmol (Evans, 1989; Al-Harbi et al., 1994), while others indicate the source of the true myrrh to be C. myrrha (e.g., FAO, 1995) and still others report that the two species are just synonymous (e.g., Vollesen, 1989; Yen, 1992). In conclusion there are several areas that research has to address related to the subject of frankincense and myrrh producing species and the vegetation resources of Ethiopia. The following is a general outline:

- One of the most difficult areas of research is to properly describe the taxonomy of the plants that produce frankincense and myrrh.
- The identification of the various oleo-gum resins collected from these species has always been a problematic area, particularly when it comes to standardizing the chemistry and potent components therein since adulteration is done most of the time or even proper sorting is absent in the areas of origin due to lack of know-how. This might have hampered discoveries of products

of much importance. Therefore, sorting individual species or sub-species gum resins and characterizing the chemistry of their gum resins is very essential.

- Survey of the distribution, densities and production potentials of the existing stands as well as possibilities for improved management and improved production systems such as tapping techniques, including artificial plantations in the natural environment of the resources for sustained productivity is very essential. This will, definitely, open more demand for these renewable and natural products.
- Detailed studies of potential markets including possibilities for creation of new local industries such as value-adding processing are needed. Local industries could be expanded at least for the already known products from the resources, while looking for further markets of reliable volume, quality requirements, etc.
- Owing to lack of adequate scientific techniques of tapping and collection procedures, it has become difficult to optimize the utilization and maintain a good quality standard of this indigenous natural product. At present almost all of the gum-resin resources at present are collected from the wild and untended plants by pastoralists, who use crude and haphazard method of incising the main stem by axes. Heavy tapping could injure the cambium and curtails the life span of the tree on account of poor wound-healing. No doubt that research supports in the areas of tapping techniques and production enhancements are very much desired.

These research efforts could contribute greatly to the production as well as economic and ecological utilization of these vast untapped renewable resources for the benefits of the local, national as well as international communities.

REFERENCES

1. Ahmed, T. (1982). The Swiss, Italian and Finnish mar-kets for Ethiopian gum Olibanum. Program for development cooperation, Market Research Report No. 5. The Helsinki School of Economics.

2. Al-Harbi M.M., Qureshi S, Raza, M., Ahmed, M.M., Giangreco, A.B. and Shah, A.H. (1994). Anticar-cinogenic effect of Commiphora molmol on solid tumors induced by Ehrlich carcinoma cells in mice. Chemotherapy 40:337–47.
3. Anonymous (1997). Technical manual for the devel-opment, propagation and production of incense resource base, Bahrdar. Bureau of Agriculture, Amhara National Regional State, Ethiopia.
4. Anonymous (2001). Tapping Ethiopia's gum arabic, www.telecome.net.et. Accessed Jan. 2002.
5. Azene Bekele-Tessema, Birnie, A. and Tengnas, B. (1993). Useful trees and Shrubs for Ethiopia. Identification, propagation, and management for agricultural and pastoral communities. Regional soil conservation Unit, SIDA, Nairobi.
6. Bhatt, J.R., Nair, M.N.B. and Mohar Ram, H.Y. (1989). Enhancement of oleo-gum resin production in Commiphora Wightii by improved tapping technique. Currett Sscience 58(7):349–357.
7. BHP (1996). British Herbal Pharmacopoeia British Herbal Medicine Association.141, Exeter, U.K.
8. Bruneton, J. (1995). Pharmacognosy, Phytochemistry, Medicinal Plants. Paris: Lavoisier Publishing.
9. Budavari, S. (ed.). (1996). The Merck Index: An Encyclo-pedia of Chemicals, Drugs, and Biologicals, 12th ed. Whitehouse Station, N.J.: Merck & Co, Inc. 1086.
10. Chickamai, B. and Gachathi, N. (1994). Gum and resin resources in Isiolo district, Kenya: Ethno-botanical and reconnaissance survey. East Af-rica Agricultura and Forestry Journal 59 (4): 345-351.
11. Coulter, J. (1987). Market study for frankincense and myrrh from Somalia (unpublished). Study un-dertaken for the European Association for Co-operation. Natural Resources Institute. Chatham, UK.
12. Council of Europe (1981). Partial agreement in the Social and Public Health Field: Flavouring sub-stances and natural sources of flavourings. List N2, no. 150. Strasbourg: Editeur Maisonneuve SA.

13. DAB (1997). Deutsches Arzneibuch. Deutscher Apotheker Verlag, Stuttgart..
14. DAB (1991). Deutsches Arzneibuch, 10th ed. Vol. 1–6. Kommentar. Wissenschaftliche Verlagsgesellschaft, Stuttgart.
15. Eden, M.J. and Parry, J.T. (1996). Land degradation in the tropics. Environmental and policy issues. Cassell Imprint, London/New York.
16. EFAP (1993). Ethiopian Forestry Action Program, Vol.(2). The Challenge for Development. Minis-try of Natural Resources Development and En-vironmental Protection, Addis Ababa.
17. Ermias Dagne, Aman Dekebo, Engida Desalegn, Tefera Bekele, Hailemichael Tesso and Daniel Bisrat (1997). Preliminary report on essential oils from frankincense, myrrh and other plants of Ethiopia. In: Conservation, Management and Utilization of Plant Gums, Resins and Essential Oils. Proceedings of a regional conference for Africa held in Nairobi, Kenya, 6-10 October, AIDGUM, GTZ.
18. Evans, W.C. (1989). Trease and Evans' Pharmacognosy, 13th ed. Bailliere Tindall, London, pp 474-475.
19. FAO (1995). Flavours and fragrances of plant Origin. Non-wood forest products 1. FAO (Food and Agricultural Organization), Rome.
20. Farah, A.Y. (1994). The milk of Boswellia forests: frankincense production among the pastoral Somali, Tia Riitta Hjort af Ornaäs (ed.). Uppsala University, Sweden.
21. Farah, A.Y. (1997). Impact of insufficient deryrain on nomad access to food in the central area of the Ethiopian Somali National Regional State. United Nations Development Programme, Emergencies Unit for Ethiopia, Addis Ababa.
22. Fichtl, R. and Admasu Adi (1994). Honeybee Flora of Ethiopia. Margraf Verlag, Weikersheim.
23. Ford, R.A., Api, A. M. and Letizia, C.S. (1992). Mono-graphs on fragrance raw materials. Food Chem Toxicol 30 (Suppl): 1S-138S.
24. Fragrance Foundation (1983). The Fragrance Founda-tion Reference Giude. Fragrance Foundation, New York.
25. Gachathi, F.N. (1997). Recent advances on classifica-tion and status of the main gum-resin produc-ing species in the family

Burseraceae. Conser-vation, management and utilization of plant gums, resins and essential oils. Proceedings of a regional conference for Africa held in Nairobi, Kenya, 6-10 October, AIDGUM, GTZ.

26. Girmay Fitwi, (2000). The status of gum arabic and resins in Ethiopia. In: Report of the meeting of the network for natural gums and resins in Af-rica (NGARA): Proceeding 29-31 May, Nairobi, Kenya.
27. Groom, N. (1981). Frankincense and Myrrh: A study of the Arabian Incense Trade. Longman, London.
28. Hillson, R.M. (1988). Gold, frankincense and myrrh. J R Soc Med 81(9): 542-543.
29. Kuchar, P. (1988). The Plants of Somalia: an Over-view and Checklist. Revised draft, LBII-USAID, CRDP, Mogadishu, 329p.
30. Kuchar, P. (1995). Identification and Characterization of Boresraceae, in the Southeastern Ethiopia. Southeastern rangelands project technical pa-per. Addis Ababa, Ethiopia, 57p.
31. Le Houérou, H.N. (1996). Climate change, drought and desertification. Journal of Arid Environments. 34(2):133-185.
32. Leung, A.Y. and Foster, S. (1996). Encyclopaedia of Common Natural Ingredients Used in Food, Drugs, and Cosmetics, 2d ed. New York: John Wiley & Sons, pp. 382–383.
33. Mielck, W. (1970). Merfluan–ein brausendes zahnsalz mit myrrhe [Merfluan–an effervescent denti-frice salt with myrrh]. Dent Dienst 22(11):21.
34. Mugah, O.J., Chicamai, B.N. Mbiru, S.S. and Casadei, E. (eds) (1997). Conservation, management and utilization of plant gums, resins and essential oils, Overview Proceedings of a regional con-ference for Africa held in Nairobi, Kenya, 6–10 October, AIDGUM, GTZ
35. Mulugeta Lemeneh, Tarekegn Abebe and Olsson, M. (2003). Gum-resins from some Acacia, Boswel-lia and Commiphora species and their eco-nomic contributions in Liban zone, Ethiopia. J Arid Environment. (In press).
36. Krishna-Murthy, T. and Shiva, M.P. (1977). Salai gug-gul from Boswellia serrata Roxb- Its exploita-tion and utilization. Indian Forester 103(7):466–474.

37. NCSS (1993). National Conservation Strategy. Vol. 1. National Policy on the Resources Base, its zation and Planning for Sustainability. National Conservation Strategy Secretariat, Addis Ababa, Ethiopia, 131 pp.

38. Provan, G.J., Gray, A.I. and Waterman, P.G. (1987). Sesquiterpenes from myrrh-type resins of some Kenyan Commiphora species. Flavour Fra-grance Journ. 2:109–113.

39. Shengyue, F. and Lihua, Z. (2001). Desertification control in China: Possible solutions. Ambio (synopsis).30 (6):384–385.

40. Steen, E. (1994). Drylands of the third world: Potential for future development (Synopsis). Ambio 23: 458-460.

41. Stiles, D. (1988). Arid Land Plants for Economic De-velopment and Desertification Control. Deserti-fication control Bulletin. 17: 18-21.

42. Tamerie Hawando (1997). Desertification in Ethio-pian highlands. RALA report No. 200. Norwe-gian Church Aid, Addis Ababa, Ethiopia.

43. Tilahun Gebremedhin (1997). Boswellia papyrifera from the Western Tigray. Opportunities, Con-straints, and Seed Germination Responses. Ethiopian M.Sc. in Forestry Prgoramme thesis works. Report No. 1996: 12. Skinnskattebergy. Sweden, p. 58.

44. Tucker, A.O. (1986). Frankincense and Myrrh. Econ. Bot. 40(4):425-433.

45. 45 Tyler, V.E. (1993). The Honest Herbal, 3rd ed. New York: Pharmaceutical Products Press.

46. Vollesen, K. (1989). Burseraceae. In: Flora of Ethiopia, Volume 3, pp. 442–478, (Hedberg, I. and Ed-wards, S., eds). National Herbarium, Addis Ababa University, Addis Abeba and Uppsala University, Uppsala.

47. Wahab, S.M.A., Aboutabl, E.A., El-Zalabani, S.M., Fouad, H.A., De Pooter, H.L. and El-Fallaha, B. (1987). The essential oil of olibanum. Planta Med 53(4): 382384.

48. Wichtl, M. and Bisset, N.G. (eds) (1994). Herbal Drugs and Phytopharmaceuticals. Medpharm Scientific Publishers, Stuttgart. Germany.

49. Wilford, J.N. (1997). Ruins in Yemeni Desert Mark Route of Frankincense Trade. New York Times. Jan. 28:B1, B10. Utilization and Planning for Sustainability.

Citations

CHAPTER 1

Zhengbin Zhang, Hongbo Shao, Ping Xu, Mengyun Hu, Weiyi Song, and Xiaojun Hu, Focus on Agricultural Biotechnology: Prospective for Bio-watersaving Theories and Their Applications in the Semi-arid and Arid Areas, ISSN 1684–5315.

CHAPTER 2

Chagonda, F. T. Mugabe, A. Munodawafa, C. P. Mubaya, P. Masere, and C. Murewi, Engaging Smallholder farmers with Seasonal Climate Forecasts for Sustainable Crop Production in Semi-arid Areas of Zimbabwe, doi: 10.5897/AJAR2014. 8509.

CHAPTER 3

Afrâa Radhouani, Mokhtar El Bekkay, and Ali Ferchichi, Effect of Substrate on Vegetative Growth, Quantitative and Qualitative Production of Muskmelon (*Cucumismelo*) Conducted in Soilless Culture, DOI: 10.5897/AJAR10.124.

CHAPTER 4

Ali Mansour Ali Saad, Noresah Mohd Shariff, and Sanjay Gairola, Nature and Causes of Land Degradation and Desertification in Libya: Need for Sustainable Land Management, DOI: 10.5897/AJB11.1235.

CHAPTER 5

James Munga Kimondo, Jacob Godfrey Agea, Clement Akais Okia, Andrew Woiso Dino, Refaat Atalla Ahmed Abohassan, Jackson Mulatya, and Zewge Teklehaimanot, Distribution and regeneration status of Vitex payos (Lour.) Merr. In Kenyan drylands, doi: 10.5897/JHF2014.0362.

CHAPTER 6

Murthy IK, Gupta M, Tomar S, Munsi M, Hegde RTGT, et al. (2013), Carbon Sequestration Potential of Agroforestry Systems in India, J Earth Sci Climate Change 4: 131. doi:10.4172/2157-7617.1000131.

CHAPTER 7

JG Annandale, RJ Stirzaker, A Singels, M van der Laan, and MC Laker, Irrigation scheduling research: South African experiences and future prospects, http://dx.doi.org/10.4314/wsa.v37i5.12.

CHAPTER 8

Mulugeta Lemenih and Demel Teketay, Frankincense and Myrrh Resources of Ethiopia: I Distribution, Production, Opportunities for Dryland Development and Research Needs, ISSN: 0379–2897.

Index